Hot Science is a series exploring t̶h̶...
and technology. With topics from̶... rewilding,
dark matter to gene editing, these are books for popular
science readers who like to go that little bit deeper ...

AVAILABLE NOW AND COMING SOON:

Hot Science series editor: Brian Clegg

THE SCIENCE OF MUSIC

How Technology Has Shaped the Evolution of an Artform

ANDREW MAY

ICON

Published in the UK and USA in 2023
by Icon Books Ltd, Omnibus Business Centre,
39–41 North Road, London N7 9DP
email: info@iconbooks.com
www.iconbooks.com

Sold in the UK, Europe and Asia
by Faber & Faber Ltd, Bloomsbury House,
74–77 Great Russell Street,
London WC1B 3DA or their agents

Distributed in the UK, Europe and Asia
by Grantham Book Services,
Trent Road, Grantham NG31 7XQ

Distributed in the USA
by Publishers Group West,
1700 Fourth Street, Berkeley, CA 94710

Distributed in Australia and New Zealand
by Allen & Unwin Pty Ltd,
PO Box 8500, 83 Alexander Street,
Crows Nest, NSW 2065

Distributed in South Africa
by Jonathan Ball, Office B4, The District,
41 Sir Lowry Road, Woodstock 7925

Distributed in India by Penguin Books India,
7th Floor, Infinity Tower – C, DLF Cyber City,
Gurgaon 122002, Haryana

Distributed in Canada by Publishers Group Canada,
76 Stafford Street, Unit 300,
Toronto, Ontario M6J 2S1

ISBN: 978-178578-991-5

Typeset in Iowan by Marie Doherty

Printed and bound in Great Britain
by Clays Ltd, Elcograf S.p.A.

ABOUT THE AUTHOR

Andrew May is a freelance writer and former scientist, with a PhD in astrophysics. He is a frequent contributor to *How It Works* magazine and the Space.com website and has written five books in Icon's Hot Science series: *Destination Mars, Cosmic Impact, Astrobiology, The Space Business* and *The Science of Music*. He lives in Somerset.

CONTENTS

MUSIC AND TECHNOLOGY 1

The idea that music and science have anything to do with each other would have struck most people as ridiculous 50 years ago. After all, music is an artform, and the arts and sciences are often seen as opposites – or at any rate, non-overlapping domains. Today, however, the gap between science and music may seem less of a gulf, thanks to the ubiquity of music and audio technology.

But technology is just the tip of the iceberg. It's the thing we're most aware of, but in fact, the relationship between science and music goes much deeper. One purpose of science is to pave the way to new technology, but it's also about improving our understanding of the world we live in. And, of course, music is an integral part of that world. Science can help to show us how music works, both in terms of the way it is created and the way we hear and respond to it. As surprising as it may seem, much of music is really quite mathematical in nature, from its basic scales and rhythms to the complex ways that different chords relate to each other. Musicians don't need a conscious understanding of these mathematical relationships – for most of them, it's a matter of intuition – but it can be interesting to look at them all the same. If nothing else, it shows that the domains of art and science aren't that far apart after all.

This studio, used by electronic music pioneer Karlheinz Stockhausen, resembles a scientific laboratory.

Some of the connections between science and music have become more visible through the use of technology, such as synthesisers and digital audio workstations (DAWs), but the fact is the connections have always been there. From a scientific perspective, all sounds, whether musical or otherwise, are vibrations in the air – or vibrations in any other medium, whether gas, liquid or solid. These vibrations spread out from the point of origin in the form of a wave, which is one of the most fundamental concepts in physics. Gravitational waves, radio waves, light rays, X-rays and even electron beams all propagate from A to B in the form of waves. In each case, something – some measurable quantity – is going up and down like a wave on the surface of the sea. In the case of a sound wave, the thing that goes up and down is pressure, so in contrast to an obviously undulating water wave, you have a moving pattern of compression and rarefaction. It's that pattern which, when it enters our ears, our cleverly evolved hearing system interprets as sound.

So how is a sound wave created in the first place? All sorts of natural processes can do that – anything that causes a disturbance in the surrounding air, from raindrops to a thunderclap. And, of course, human vocal cords. That's how music started – with simple a cappella singing – but to take it beyond that, we have to turn to technology. The *Concise Oxford Dictionary* defines technology as 'the application of scientific knowledge for practical purposes', so we're not necessarily talking about electronics here. We'll see in the next chapter how the ancient Greeks worked out how the design of a wind, string or percussion instrument determines the shape of the sound waves it produces. In that sense, even a drum, flute or lyre is 'music technology'.

Another big impact that technology has had on music is the ability to record a performance and play it back later. Today, that's inextricably linked to electronics, but it doesn't have to be. The first practical phonograph, invented by Thomas Edison in 1877, was purely mechanical. An ear trumpet-like horn amplified incoming sounds sufficiently to vibrate a stylus, which made a copy of the audio waveform onto a wax cylinder that was rotated at a steady rate by a clockwork mechanism. Playing back the sound was literally the reverse of recording: the rotating cylinder made the stylus vibrate and the horn amplified the sound to the point that it was audible.

A simple contraption like this is never going to produce the kind of high-fidelity sound you could sit and listen to with pleasure, but it's good enough to make some kind of permanent record of a performance. And that's where we come to the real irony – or maybe even tragedy – of the situation. Edison's phonograph was so simple it could have been made, if it had crossed anyone's mind to do so, centuries earlier. So just think of all the musical performances that could have been recorded but weren't. Both Bach and Beethoven were popular keyboard players as well as composers – and their real crowd-pleasers were virtuoso improvisations rather than written-out compositions. That, however, was in the days before phonographic recording, so all those performances are lost to history.

In fact, in the case of a keyboard player, such a recording doesn't even require a phonograph. Sometime after Edison's invention, someone belatedly worked out that, using a mechanism inside a specially modified piano, you can record a pianist's performance on a roll of perforated paper and then play it back automatically as often as you like. The 'player piano', as it was called, was very popular between its invention in 1896 and the advent of good (or at any rate acceptable) quality phonograph recordings in the 1920s, after which it fell into obscurity. But maybe not completely so. If you've ever used a DAW, you'll be familiar with its 'piano roll' editor – a name that harks back to the old perforated paper rolls of a player piano.

There's a rather silly trend in classical music these days towards 'authenticity' – getting as close as possible to the notated score, which is presumed to represent the composer's exact intention. But that might not be the case at all. When Chopin played one of his mazurkas – a Polish-style dance notated with three beats in a bar – for the conductor Charles Hallé, he played it with four beats in a bar. When Hallé pointed this out, Chopin explained that he was reflecting the 'national character of the dance'. If that performance had been captured on a piano roll or wax cylinder, Chopin's mazurkas might be played in a completely different way today.

As it happens, we do have a recording of one famous 19th-century composer playing the piano. Johannes Brahms died in 1897 – which happens to be the same year that a Cambridge professor named J.J. Thomson discovered the electron, so you know the quality of recordings made at this time are not going to be good. Made in 1889 using an Edison phonograph, it's a strong contender for the worst recording of a great musician ever made. You can hear it in a YouTube video made by a talented young pianist calling himself 'MusicJames'.* It's well worth watching the whole thing; with the aid of some clever detective work, James disentangles exactly what Brahms is playing – and it's not what any classical pianist today would expect.

The basic problem with the recording becomes apparent right at the start when Brahms speaks a few words in an unnaturally high-pitched voice. That's because the wax cylinder only preserved higher frequency sounds, so that when he starts playing one of his Hungarian dances, you can only hear the notes his right hand plays, not the left. The trouble, from a musician's point of view, is that what the right hand is playing doesn't quite match the notated score. By interpolating

---

* https://www.youtube.com/watch?v=XEDGPG8C5Y4

what must be going on in the left hand, James concludes that Brahms is playing in what might well have been the standard pianistic style of his time, full of 'rhythmic improvisation techniques which have gone extinct in classical music over the last century'.

Without technology, even in the crude form of an Edison phonograph, musical insights like this would be impossible. And without technology in the far more sophisticated form of digital workstations, electronic synthesisers and streaming services, a vast proportion of today's music simply wouldn't exist. In turn, the technology itself wouldn't exist if someone – generally the developers rather than the users – did not have a strong grasp of the underpinning science.

As an example, consider the brilliant, genre-hopping musician Jacob Collier, who is still in his twenties as this book is being written. There's no obvious science in his background or upbringing. His mother is a music teacher, violinist and conductor, and his grandfather was leader of the Bournemouth Symphony Orchestra. As a child, Collier sang the key role of Miles in three different productions of Benjamin Britten's *The Turn of the Screw*, one of the greatest of all 20th-century operas. In 2021, his rap song 'He Won't Hold You' won a Grammy award, and he narrowly missed getting Album of the Year too. Part of his popularity undoubtedly comes from a series of YouTube videos in which he shows exactly how he creates his music. And now we come to the point, because some of his explanations look more like science than art.

Like many musicians today, Collier puts his songs together almost entirely with the aid of software, in the form of a DAW. Apple users, whether they're musicians or not, probably have some knowledge of how this works through playing around with the GarageBand DAW that comes free with Mac computers and iPads and iPhones (albeit in cut-down form). The DAW used by Collier is a more sophisticated Apple product called Logic Pro, but from a functional point of view it's very similar to GarageBand, as are numerous other DAWs made by other companies. The user interface is oriented towards musicians, but the back end of the software is pure science – specifically, the science of acoustics and audio signal processing. This occasionally breaks through in Collier's videos, when he uses distinctly scientific-sounding jargon like 'waveform' and 'quantisation'. Albert Einstein would have understood those terms, but a musician of his era, such as Ravel or Rachmaninov, probably wouldn't.

Another multi-talented young musician is Claire Boucher, who creates music under the name Grimes. Her music is even more intensely electronic than Collier's. As she put it in a 2015 interview: 'I spend all day looking at fucking graphs and EQs and doing really technical work.' EQ, of course, means equalisation, and to many people it's simply another piece of music jargon. But – as we'll see in the chapter on electronic music – it really means changing the shape of an audio signal in the frequency domain. EQ has a lot more to do with physics than music theory.

The use of digital technology has transformed the whole sound of contemporary music. In the hands of a great artist/producer, the act of sampling snippets of existing works – anything from a well-known James Bond theme to obscure Eastern European rock music – has become an art form in itself. Through skilful manipulation of samples – looping, re-pitching, speeding up or slowing down and adding a host of subtle electronic effects – it's possible to add whole new layers of emotional texture to a song in a way that would have been impossibly time-consuming a few decades ago.

It's a similar story across all music genres. Classical composer Emily Howard has a university degree, not in music, but in mathematics and computer science – both subjects she frequently draws on in her work. In the world of electronic dance music, Richard D. James is best known by the alias of Aphex Twin, but one of his early albums was released under a different pseudonym, Polygon Window. Its title, *Surfing on Sine Waves*, is an electronic musician's pun – yet there was a time, not so long ago, when no one but an engineer or scientist was likely to know what a sine wave is. When the 'avant-pop' artist Björk released her heavily electronic album *Biophilia* in 2011, it was launched in parallel with a mobile app that further explored the musical and thematic background of the songs. So, in numerous different ways, music and technology are becoming inseparable in the modern world.

Science has had an impact on the way people listen to music too. If you use a streaming service, you're probably concerned about having sufficient bandwidth – another term that, until a few decades ago, was virtually unknown outside the sciences. Or maybe you use noise-cancelling headphones, which are entirely reliant on the physics of sound waves for their effectiveness. Even the very word 'electronics' – inextricably associated with music technology in many people's minds – comes from the electron, one of the seventeen elementary particles of modern physics. Without a thorough understanding of its behaviour, the electronic technology most of us depend on for our music would simply be impossible.

All this technology is rooted, as the *Oxford Dictionary* says, in basic science. That's not to say, of course, that musicians have to *understand* the science in order to exploit the technology; fortunately, it's quite the opposite. While most scientists (like many people in all walks of life) are music lovers, not all musicians are science lovers.

That's always been the case. In 1877, the British physicist Lord Rayleigh produced a 500-page textbook called *The Theory of Sound*, which became a classic of its kind in the scientific world. But not in the musical world. Towards the end of his life, the great composer Igor Stravinsky admitted: 'I once tried to read Rayleigh's *Theory of Sound* but was unable mathematically to follow its simplest explanations.'

If you find yourself sympathising with Stravinsky, you'll be pleased to hear there isn't much in the way of textbook physics or complicated mathematics in this book. I just want to highlight a few of the many respects in which music is more closely related to science than it might appear – in a way that I hope is both thought-provoking and entertaining. Here's a taste of what's in store:

Chapter 2 takes a closer look at sound – musical and otherwise – from a scientific point of view. This is where we really get to grips with the concept of sound as a wave or vibration and trace the history of this notion all the way from ancient Greece to the present day. It's in this chapter that some of the sciencey-sounding music jargon (or musicky-sounding science jargon?) that I've already bandied about, such as sine waves and the frequency domain, get a proper explanation.

Chapter 3 is all about musical parameters and algorithms – but it hasn't got anything to do with computers. Musicians have been playing with 'parameters' like the pitch and duration of notes and applying 'algorithms' – which ultimately are just collections of rules – to the way they string notes together since time immemorial. And by the middle of the 20th century, with composers like Arnold Schoenberg, John Cage and Karlheinz Stockhausen on the scene, musical algorithms started to take on a distinctly 'scientific' appearance.

To most people, if the term 'scientific music' means anything at all, it refers to modern electronic music – and that's the subject of Chapter 4. This takes us all the way from the birth of electronics in the early 20th century, through the pioneering experiments of Cage, Stockhausen and others in the 1950s, to the ever-widening use of synthesisers through the 1960s, '70s, '80s and '90s – before finally coming up to date in today's world of DAWs, plugins and computer algorithms.

That might be the end of the book if we were just talking about music technology, but actually, there's a whole other side to the science of music, and that's the 'brain science' behind how we perceive music, and how musicians create it. This is a topic we look at in the final chapter, together with the intriguing question of whether, as computers get more and more powerful, they will ever be able to create their own music on a par with that of human musicians.

While writing the book, there were various points – particularly when discussing science-inspired composition methods – where I couldn't resist trying my hand at it. With the caveat that I'm just a computer geek, not a musician, you can check out the results on a YouTube playlist I created, entitled 'The Science of Music'.

THE PHYSICS OF SOUND 2

In space, as everybody knows, no one can hear you scream. That's because space is a vacuum, and a sound wave – which is basically a rapid sequence of pressure fluctuations – needs a material substance to travel through. It doesn't have to be a particularly dense substance, though. The atmosphere of Mars, at ground level, is almost 100 times thinner than our own, but it's still enough to carry sound waves. We know that for certain now, because NASA's Perseverance rover, which landed on the red planet in February 2021, has a couple of microphones on board. NASA has a great web page where you can hear some of the sounds they've recorded, including gusts of Martian wind, the Ingenuity helicopter in flight and the trundling of the rover's own wheels.*

The same web page also includes a selection of audio clips recorded on Earth processed to sound the way they

---

* 'Sounds of Mars', NASA, https://mars.nasa.gov/mars2020/participate/sounds/

would if they were played on Mars. Compared with the originals, these are fainter and more muffled-sounding, with higher-pitched sounds like the tweeting of birds being almost inaudible. That's due to the different properties of the Martian and terrestrial atmospheres, which affect the way sounds propagate through them. All in all, though, it's surprising – and maybe a little disappointing – just how ordinary the Martian sounds recorded by Perseverance are.

Coming back down to Earth, you may have noticed – if you've spent any time snorkelling in the sea, for example – that sound travels through water as well as air. In fact, it generally carries further underwater than through the atmosphere; as you drift above a reef, a jet ski half a kilometre away may sound like it's about to crash into you. That's why whales, like humans, use sound to communicate. Some whale songs can remain audible over hundreds or even thousands of kilometres.

In the human world, navies use underwater sound waves, in the form of sonar, to hunt for things like enemy submarines. This is a choice that's forced on them by the laws of physics because radio waves don't really work underwater – they're attenuated to virtually nothing within a few metres – so the more obvious choice of radar is a non-starter. But pretty much everything that a radar can do above water – such as determining the distance, speed or even shape of a target – can be achieved equally well in the underwater domain by sonar. And the analogy between radio and sound waves is a useful one, because – from a scientific point of view, at least – a radio wave may be a more familiar concept to many people. We can use it to explore just how sound waves work in a bit more depth.

Measuring sound waves

Two common terms in the context of radio are 'frequency' and 'wavelength'. FM radio stations, for example, generally identify themselves by the frequency they're broadcasting on – X megahertz, say – while older stations would often specify a wavelength in metres. The two numbers aren't independent – if you know one, you can work out the other. All radio signals travel through the Earth's atmosphere at the speed of light, approximately 300 million metres per second. If you imagine a cyclic wave pattern travelling at that speed and repeating itself once every metre – having a wavelength of one metre, in other words – then it will whiz past you at a frequency of 300 million cycles per second. The technical shorthand for 'cycles per second' is hertz (Hz), so putting all that together, the frequency of the wave is 300 megahertz.

Sound waves work in much the same way, albeit travelling at a much slower speed. The speed of sound in a given medium depends on a variety of factors, including pressure, temperature and composition. Under normal conditions at the Earth's surface, it's around 340 metres per second, but on Mars it's closer to 240 metres per second. On the other hand, the speed of sound in water is much faster – typically around 1,500 metres per second. So sound waves don't always have the same one-to-one relationship between frequency and wavelength as radio. Yet sounds in different environments, such as underwater or on Mars, don't really sound all that different to our ears. It turns out that the really important property of a sound wave is its frequency, which determines the rate at which our eardrums vibrate and hence the sound our brains hear. So, for most purposes, we might as well forget about the speed and wavelength of sound waves and just concentrate on their frequency.

Sound frequencies are generally much lower than the frequencies used by radio stations, ranging up to a few tens of kilohertz (kHz) rather than hundreds of megahertz (MHz). Technically, we could say that, for example, an audio broadcast has a 'carrier frequency' of 100 MHz, but a *bandwidth* – i.e. the amount of useful information it contains – of 50 kHz. It's easy enough to see why this has to be the case. The radio wave can only carry an audio signal if its shape is varied – or 'modulated', to use the technical term – in sync with the sound wave that's being transmitted, but obviously, this modulation can't change the frequency of the radio signal very much because it has to stay within the channel width of a suitably tuned radio receiver.

The upshot is that the carrier frequency of the radio wave has to be much higher than the frequency of the modulating audio signal that it's conveying. You may think I'm being terribly technical here, but the fact is you've already heard about the principle I'm talking about. It's in the names of the two different types of radio broadcasting: amplitude modulation (AM) and frequency modulation (FM). Of the two, AM is the easiest to visualise, as shown in the diagram opposite. FM is harder to illustrate because it involves changing the frequency of the wave (affecting the left–right distance between peaks) rather than the amplitude (up–down distance between peaks).

The human ear has an effective range between around 20 Hz and 20 kHz, although the upper frequencies tend to disappear in those of us of more mature years as the associated sensory cells atrophy (no one seems to know why this happens, but speaking for myself, I never thought those high

The process of amplitude modulation, showing (top) an unmodulated radio wave, (middle) an audio signal and (bottom) the modulated radio wave; in reality, the difference between the radio and audio frequencies is much greater than shown here.

frequencies were much use anyway). As a general rule, we perceive low frequencies as low-pitched sounds and high frequencies as high-pitched ones – a relationship that seems simple enough, but actually isn't quite as straightforwardly linear as you might imagine. We'll come back to it before very long, but there's another basic issue of measurement that's worth covering first.

Going back to the radio analogy, another important property of a radio wave is its power; obviously the more power a transmitter produces, the easier it is to pick up a signal. In the radio context, as with electronics in general, we usually measure power in watts or kilowatts – or, in the case of the tiny amount of power entering the receiving antenna of a radio, microwatts or nanowatts. I'm assuming everyone is familiar with these prefixes I'm using: mega for a million, kilo for a thousand, micro for a millionth and nano for a billionth. They're in pretty common use in all walks of life. But there's one prefix that's fallen into almost complete disuse, and that's deci for a tenth. In fact, the only deci-unit I can think of is the decibel, which happens to be a measure of power that's most commonly encountered in the context of sound. And it may be the weirdest unit of all.

A hundred watts is twice 50 watts. That's so blindingly obvious I'm embarrassed to have to say it. A hundred of virtually any measurement unit you can think of is twice 50 of the same thing. But there's an exception to every rule, and in this case the exception is the decibel. A hundred decibels is not twice 50 decibels – it's *100,000 times* 50 decibels. How on earth did that happen? Fasten your seatbelt because this is going to need a little mathematics.

We might reasonably suspect that a decibel is a tenth of a bel, but what exactly is a bel? It comes from a mathematical function you may have noticed on some calculators called a logarithm, or 'log' for short. This is essentially the opposite of a 'power', in the sense that 10 to the power of 2 is 100, or 10 to the power of 3 is 1,000. So the log (to the base 10) of 100 is 2, and the log (to the base 10) of 1,000 is 3. If you've followed me so far, then we've almost got to the definition of a bel, which is the log to the base 10 of the ratio of two sound levels.*

---

* You may wonder what happens when the ratio doesn't come out as a power of ten. But in fact, *all* numbers can be expressed as powers of ten, if you include fractional and negative powers.

To complicate matters a little bit more, no one ever actually talks about bels – they multiply the number by ten to give the result in decibels.

You've probably heard decibels mentioned countless times in the context of loud sounds, and if you're into music software, you'll be used to thinking of relative sound levels in terms of 'dB'. But now you may be wondering *why* physicists and sound engineers inflicted such a mathematically evil unit on us. The reason is that it reflects the way we actually hear sounds; our ears tend to work logarithmically rather than linearly. A step up of 10 dB is equivalent to multiplying the power, as measured in watts, by a factor of ten. That's why I said earlier that 100 dB is 100,000 times 50 dB. But the fact is that we *perceive* each 10 dB step as more or less equal. Subjectively, going from 90 to 100 dB sounds much the same as going from 80 to 90 dB – even though objectively the difference in power is ten times greater.

One result of this is that we tend to underestimate just how much power there is in really loud sounds – a misperception that can become quite dangerous if we get too blasé. By convention, 0 dB is taken as the 'threshold of hearing' – the quietest sound the human ear can perceive. On this scale, the ambient noise level in a typical suburban street might be around 50 dB, while potential issues start at just over 100 dB – things like motorbikes, pneumatic drills and car horns. Close proximity to such sounds can lead to hearing damage and other health effects – a fact that, sadly, hasn't gone unnoticed by the military. There's a whole new generation of sonic weapons, such as the long-range acoustic device (LRAD), which deliberately projects harmful levels of sound over a distance of 60 metres or more. At its upper end, the LRAD functions around 160 dB (that's 10^6 or a million times 100 dB, remember), which can cause damage to internal organs when used at close range.

'Loud noises don't bother me,' you might say, 'because I've got a pair of noise-cancelling headphones.' Actually, I'm glad you mentioned that because it reminds me of another bit of wave physics I wanted to talk about. It's a basic property shared by all types of waves called 'interference'. That's a technical term, but it means pretty much what it says. If two waves with different patterns of high and low intensity try to occupy the same space, they interfere with each other. That isn't always a good thing, but it can be if it's exploited cleverly enough. It's the basic idea behind noise-cancelling headphones, and it sounds a bit like black magic even though it's just physics. One of the best explanations I've seen of the principle comes from a sci-fi story, 'Silence Please', which Arthur C. Clarke wrote in 1950:

> If one could get two sets of waves exactly out of step, the total result would be precisely zero. The compression pulse of one sound wave would be on top of the rarefactions of another. Net result – no change and hence no sound.

While the basic idea was well known to scientists at the time of Clarke's story, actually putting it into practice was a different matter. It requires a microphone to detect the sound wave to be cancelled and a complicated electronic circuit – suitably miniaturised – to produce a precisely inverted version of that wave. So, it was only in the 1980s, decades after Clarke's story, that noise cancellation became a practical reality.

Musical acoustics

Now that we've covered the basics of sound waves, we can look in more detail at how these apply to music. Let's start by rewinding a few thousand years. The first person to make anything approaching a scientific study of music was the Greek philosopher Pythagoras, all the way back in the 6th century BCE. He's best remembered today for his theorem about right-angled triangles – i.e. that the square on the longest side is equal to the sum of the squares on the other two sides (a fact which, surprisingly enough, we'll come back to shortly in the context of music). As one of the foundations of mathematics, you might think this makes Pythagoras eminently qualified as a protoscientist – but actually, he wouldn't have thought of himself that way at all.

The close link between mathematics and science lay the future, starting with the work of people like Archimedes three centuries later. Before that, the physical world was seen as random and capricious, with the ordered elegance of mathematics reserved for higher planes of existence. That was how Pythagoras saw it, and his philosophy – and the mathematics wrapped up in it – was much less concerned with the physical world than the mystical one. Like his approximate contemporary Gautama Buddha, for example, he believed in a cycle of rebirth from one life to another. And just like Buddha, we don't know very much about Pythagoras beyond the anecdotes and legends handed down by his followers.

According to one of these legends, Pythagoras was passing a group of blacksmiths one day when he observed that the clanging of some of the hammers was more 'in tune' than others. On closer investigation, he discovered this was due to the different sizes of the hammers – which prompted him to undertake a methodical study of the way objects produce musical sounds. Regardless of whether the blacksmith story actually happened, Pythagoras does seem to have carried out a series of practical experiments using a simple string instrument which showed the basic mathematical nature of music for the first time. Ironically, Pythagoras himself – in keeping with his general philosophy – took this as proof that music was mystical rather than earthly in nature.

What Pythagoras essentially discovered was that there are simple relationships between the length, weight and tension of a vibrating string and the sound it makes. If you've ever studied a guitar – in the sense of investigating it for yourself, I mean, rather than being shown how to use it by someone else, which is cheating – you'll have noticed a few basic things about it. It has a number of strings (usually six) of roughly equal length but different thickness and different tightness. They all make different sounds when plucked, but there don't seem to be enough of them to play a proper tune (except maybe the intro to Metallica's 'Nothing Else Matters', which has only four different notes, all corresponding to standard open guitar string tunings). The real breakthrough comes when you notice the metal bars, or frets, across the neck of the guitar. By putting your fingers on these, you can shorten the effective string length and play any note you want to.

This was basically the process Pythagoras went through all those centuries ago. He found the shorter he made a string, the higher the pitch of the sound it produced. On top of that, there was a gratifyingly mathematical relationship between length and pitch, with the most harmonious-sounding pitch intervals corresponding to the simplest string–length ratios, such as 2:1 or 3:2. In the terminology of today, those

**A 15th-century woodcut showing Pythagoras
experimenting with a musical instrument.**
Public domain image

particular intervals are called an 'octave' and a 'fifth', based
on their positions in one specific musical scale that's common in the West. From a scientific point of view (or in the
context of music from other parts of the world) these terms
are really very misleading, because the intervals in question have no fundamental connection to the numbers eight
and five at all. However, we're stuck with them. I'll leave a
detailed discussion of how scales are constructed to the next
chapter; for now, I'll just mention how the basic concept can
be traced back to Pythagoras.

   As far as our ears are concerned, two pitches an octave
apart, such as 'middle C' on a piano and the next C up,

represent two instances of the same note – or the same pitch class (PC), to use the proper jargon. But a pitch a fifth above that PC – G in this case – is a different PC, while going up another fifth to D takes us to yet another PC, and so on. Pythagoras noticed that after you've done this twelve times, you're pretty much back to the PC you first thought of. Actually, he seems to have believed – through mystically inspired wishful thinking, probably – that going up in fifths twelve times takes you *exactly* back to the original PC. Actually, it doesn't; by making that assumption, you end up with a slightly wonky scale called 'Pythagorean tuning'.* On the other hand, the 'equal temperament' tuning found on modern pianos and MIDI keyboards, which divides an octave into twelve equally spaced intervals called 'semitones', doesn't produce the neat cycle of *exact* fifths that Pythagoras wanted. But as compromises go, it's a pretty good one – and it's the one almost everyone uses today.

From a physics point of view, it's not too difficult to work out how a string instrument manages to make a musical sound. The same kind of principles apply to wind instruments, as well as percussion instruments, such as drums or Pythagoras' blacksmith hammers. However, strings are that bit simpler, so we'll stick with them for the moment. When a string is plucked, it vibrates in a wave-like way, but because the ends are fixed, only certain wavelengths are going to fit. The simplest mode of vibration, referred to as the 'fundamental', is the one where *exactly half a wavelength* fits into the string length. The maximum displacement, or 'amplitude' in scientific jargon, occurs in the exact middle of the string. By

---

* I'll put this in more scientific terms later, after I've said a bit more about how sound waves work.

disturbing the air around it, this vibration creates a sound wave with the same frequency as the vibration frequency of the string.

So, what is that frequency, then? We already know the wavelength of the vibrating string because that's twice the length of the string itself (as I just said, exactly half a wavelength fits into the string length), so we just have to divide this wavelength into the speed with which a wave travels along the string. This is determined by two factors that, from playing around with a guitar, we already know the sound depends on: the tension of the string and its weight per unit length. Specifically, the speed of the wave is the square root of the ratio of those two numbers – a neat relationship that you might expect me to tell you was discovered by some brilliant early scientist like Galileo. In fact, it was discovered by that particular brilliant scientist's dad, Vincenzo Galilei, who happened to be a musician – and clearly a pretty good amateur scientist himself.

Galileo reciprocated by making an interesting discovery about music of his own. Pythagoras had worked out that harmonious pitch intervals correspond to simple wavelength ratios like 2:1 or 3:2, and that the more complicated the ratio, the less 'harmonious' the resulting sound. Taking this observation to an extreme, Galileo worked out what the least harmonious ratio of all must be. His argument is going to take a little unpicking, so we'll start with a quotation from Galileo himself:

Especially harsh is the dissonance between notes whose frequencies are incommensurable. Such a case occurs when one has two strings in unison and sounds one of them open, together with a part of the other which bears the

same ratio to its whole length as the side of a square bears
to the diagonal; this yields a dissonance similar to the tritone or semi-octave.

There's quite a lot of science in that paragraph, so we'll take
the ideas one at a time. When you hear the word 'ratio' you
tend to think of the ratio between two whole numbers, the
simplest of all being 2:1. When applied to the frequencies
of sound waves, we already know this corresponds to the
musical interval called an octave – for example, from the
A below middle C (frequency 220 Hz) up to the A above
middle C (440 Hz). But what's half that – a 'semi-octave',
as Galileo calls it? The answer isn't as obvious as you might
think because the relationship between (scientific) frequency and (musical) pitch isn't a linear one. A frequency of
440 Hz is an octave above 220 Hz but an octave above 440 Hz
is 880 Hz, not 660 Hz. To our ears, the step from 440 Hz to
880 Hz sounds the same as the step from 220 to 440 Hz
– they're all 'A' to us – even though numerically it's twice
as big.

It's like that decibel nightmare all over again. You'll
remember that decibels are derived from the logarithm of an
acoustic power ratio, because our ears happen to hear sound
intensities on a logarithmic scale. In the same way, we hear
frequencies logarithmically too, and what we call musical
pitch is proportional to the *logarithm* of the frequency. In
the context of decibels, we talked about logarithms as being
powers of ten, but they can be powers of any number we
want to use as the 'base' of our logarithms. In the case of
pitches, it's easiest to take two as our base. The number
of octaves represented by a pitch ratio of 2^n is n. In the case of
a musical fifth, for example, the power of two corresponding

to a ratio of 1.5 is (as near as makes no difference) 7/12 – so a fifth is 7/12 of an octave.

So now we can go back to that Galileo quote and understand it a bit better. What he's looking for is some ratio that gives you exactly half an octave. Applying what we've just learned, the ratio we want must be $2^{0.5}$, which is a mathematician's way of saying the square root of two. Somewhat cryptically, Galileo referred to this as the same ratio 'as the side of a square bears to the diagonal'. We can go back to Pythagoras to help us with this. Two sides of a square and its diagonal form a right-angled triangle, so we know that the square of the diagonal must be equal to the sum of the squares of the other two sides. If we define the sides as having a length of one unit, then the diagonal must be the square root of two – exactly the ratio we're looking for.

What's special about this particular ratio? Since it's exactly half an octave, you might think it would have a nice, neat numerical value. But it's much more complicated than that because it can't be expressed as the ratio of two whole numbers at all. The square root of two is approximately 1.41421356237, but really the digits go on for ever, without ever repeating. That's what Galileo means by 'notes whose frequencies are incommensurable'; there's no way you can make a whole number of waves of one frequency fit into the same length of string as a whole number of waves of the other frequency.

If we go back to the octave example we referred to earlier, between the A at 220 Hz and the one at 440 Hz, then the half-octave point between the two is 220 × √2 = 311.127 Hz, corresponding to the E♭ above middle C. Now, if you look at a piano keyboard, you'll find that E♭ is the sixth key above A – an interval musicians refer to as six semitones or three whole tones. So that explains another strange word Galileo used when he referred to this half-octave interval as a 'tritone'.

But is the tritone really as 'harsh' as Galileo claims? If you have a MIDI keyboard to hand (either a physical one or a virtual one courtesy of a mobile app), try playing two notes six semitones apart and see what you think. Played one after the other – in either order, because it's the same half-octave interval either up or down – it sounds distinctly ominous, but not really unpleasant. Played simultaneously, the tritone sounds unstable – like you want to raise or lower one of the notes by a semitone – but again, not necessarily unpleasant.

Because of its rather sinister sound, you probably wouldn't want to put the tritone in a hymn or other religious work. The European churches felt the same way throughout the Middle Ages, sternly discouraging their musicians from using this particular interval. This gained the tritone the nickname 'the Devil's chord' – though probably only because it was never heard in church, rather than because people really believed it had satanic connections. By the 19th and 20th centuries, however – when the word satanic was turning into a synonym for 'cool' in certain circles – the association didn't do the tritone any harm at all.

Perhaps the most important development in 19th-century music was the realisation that you can be a great composer without following anyone else's rules. You can do whatever you like, so long as it produces the *sound* you're looking for. Franz Liszt was at the forefront of this movement – and it's fair to say that he loved the sound of the Devil's chord. He used it, for example, to depict hell in the opening bars of his 'Dante Sonata', and to represent a sneaky demon in his

Mephisto Waltz No. 2. Another composer of the same generation, Hector Berlioz, made prolific use of the tritone in the section of his *Symphonie Fantastique* called 'Dream of a Witches' Sabbath'.

Even if the tritone didn't have any particularly evil or devilish connotations in the Middle Ages, thanks to people like Liszt and Berlioz it does now. If you did try playing around with the tritone on a MIDI keyboard, then – in descending form – it may have recalled another 'sabbath' to mind. The eponymous 1970 song by Black Sabbath starts with exactly that motif.

So, we've established that the tritone sounds creepy, but does it really deserve that word 'dissonant' that Galileo applied to it? The problem is, it's a word with two different meanings – a deeply ingrained everyday one, and a more subtle technical one. In the ordinary day-to-day sense, a dissonant sound is an excruciatingly unpleasant one, which you'd readily commit murder to put an end to – like fingernails scraping down a chalkboard. No one is going to describe their favourite song as 'dissonant', except maybe fans of extreme metal bands like Meshuggah or Deathspell Omega. Going by the technical definition, however, it's likely that *everyone*'s favourite song contains dissonance. Not only that, but the dissonances almost certainly coincide with the song's emotional high points.

Technically, dissonant notes are ones where the corresponding frequencies aren't closely related to each other. The tritone is one example – and as I said before, it tends to sound 'unstable', as if you want it to change to something else. But that isn't a bad thing – it adds tension and movement to music. Another archetypal dissonance is when you have two notes exactly one semitone apart – or else eleven semitones, just short of the twelve that would make an octave. This sounds rather as if the wrong note has been played – but in some circumstances, the 'wrong' note can sound a lot better than the 'right' one. Think of the piano introduction to John Lennon's song 'Imagine', for example. The note on the last beat of the first bar has a slightly 'wrong' sound because it's a semitone lower than you might expect. But if Lennon had played the note a semitone higher, the result would have sounded so insipid as to be the *real* wrong note.

Nevertheless, the fact remains that, in the technical sense, the note Lennon plays is a dissonance. You can find similar dissonances scattered through virtually any piece of music; Lennon's fellow ex-Beatle Paul McCartney put a very obvious one in the opening chord of 'My Love', for example. You could even argue that the true art of writing music is knowing exactly where to put these 'wrong' notes – and how to follow a 'wrong' note with a 'right' one, or vice versa. That's true even of great classical composers like Mozart, whose name is virtually synonymous with faultlessly elegant music. The opening bars of his *String Quartet No. 19*, written in the early 1780s, are almost non-stop dissonance – to the extent that it's known by the nickname of 'The Dissonance Quartet'. Personally, I think that's very unfair, because it's anything but dissonant in the *colloquial* sense of the word. If you had to apply an adjective to it, you might say 'spiritual' or 'mysterious', but certainly not 'dissonant'.

On the other hand, depending on your musical tastes, you might describe a heavy metal band playing a series of massively distorted power chords as 'dissonant'. Yet a power chord is simply made up of octaves and fifths, so from a technical point of view it contains no dissonance at all. There's clearly more to the way musical sound works than we've explored so far. To go deeper, we'll need to look more directly at what's going on inside a sound wave.

The frequency domain

From a scientific point of view, the simplest way to visualise any wave-like phenomenon is in the form of a 'time series'. That's a technical term, but it's as good a description as any of what we're talking about: essentially a long sequence of measurements of some physical property as it changes over time. In the case of a sound wave, the property in question might be the air pressure inside our ears – or more practically, the voltage inside an electronic circuit. But a time series has its problems. Although it's a nice direct representation of what's going on, it can easily become much too complex to analyse in a meaningful way.

Sound waves come in a wide variety of shapes and sizes – as many as there are sounds themselves, in fact. One particularly interesting waveform, from a mathematical perspective, is called a sine wave – a term we briefly encountered in the first chapter. Visually this takes the form of a regularly undulating pattern, similar to the audio signal shown in the 'amplitude modulation' diagram on page 15. But its unique mathematical interest comes from the fact that it's the *only* waveform with a single clearly defined frequency.

If you've ever played with a synthesiser (or a software synth on your laptop or tablet), then you'll know that sine waves are one of the basic waveforms it can produce, along with other types such as sawtooth and square waves. But unlike sine waves, these don't have a single unique frequency. They may look simple enough when plotted as a time series, but they sound more complex – and more interesting – when played. The reason is all those extra frequencies they contain. It's the same with 'real' musical sounds, such as human voices or the instruments of an orchestra, which can sound different to our ears even when they're playing the same note. We need to understand why this is, but it's not something we can analyse simply by looking at a time series.

The person to turn to is a French physicist named Joseph Fourier and the research he conducted in the early 1820s. He discovered that it's possible to represent any given time series as the sum of appropriately weighted sine waves. That may sound a little dry and mathematical, but it provided scientists and engineers with one of the most useful analytical tools ever invented. 'Fourier analysis', as it's called, has found applications in everything from radar systems and image processing to crystallography, quantum physics and economics. Its relevance to the physics of sound was recognised early on, although that wasn't Fourier's own area of research. He was more interested in what he called the 'analytic theory of heat', and in fact, his other great contribution to science was the discovery of the greenhouse effect.*

In the context of sound, Fourier's discovery allows us to transform an audio signal from the time domain into the 'frequency domain'. That's because each of the component sine waves has a precisely defined frequency, so we can plot a graph of sound intensity versus frequency instead of versus time. Being able to visualise sound in the frequency domain in this way has always been useful to scientists, but these days it's equally useful to musicians because it enables them

---

* That's right – people understood the greenhouse effect *before* they embarked on the Industrial Revolution.

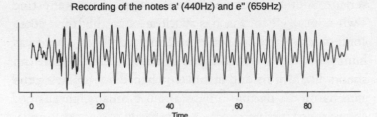

Recording of the notes a' (440Hz) and e" (659Hz)

Signal that was Fourier-transformed (frequency spectrum)

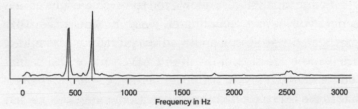

Fourier analysis can separate two simultaneous tones in the time domain (top) into two distinct peaks in the frequency domain (bottom).

to manipulate sounds in ways that would be impossible otherwise.

When I was talking about Pythagoras' experiments with musical instruments earlier, I gave the impression that a musical sound – produced by a vibrating string, for example – is a single frequency. That was really an oversimplification – although I did cover myself by referring to the 'fundamental' mode of vibration. This is the lowest frequency in a musical note, and usually the strongest. But in reality, there's a whole series of higher frequency components, called partials.

In acoustic instruments of the wind and string families, all the strongest partials – called harmonics – are multiples of the fundamental. That's why these instruments tend to have a 'pure' sound. On the other hand, an electric guitar played with a lot of distortion has much stronger high-frequency harmonics, which is why even a simple power chord can start to sound dissonant. And untuned percussion instruments like drums and cymbals produce a whole range of non-harmonic partials, giving them a much richer sound than a tuned instrument. As for the world of electronic music – well, the sky's the limit. You can tailor your own sounds with whatever arrangement of partials you like.

Fourier analysis also allows you to 'reverse-engineer' any sound you have a recording of. Jason Brown, a Canadian mathematics professor, produced an entertaining example of this on the occasion of the 40th anniversary of The Beatles film *A Hard Day's Night* in 2004. The iconic opening chord of the film's title track had always been something of a mystery to Beatles fans. Although its basic nature is clear enough, there have been perennial arguments over exactly which instruments were playing which notes.

Brown ran a Fourier analysis to identify the 48 strongest frequencies and their corresponding amplitudes. But it's not as simple as saying that all those 48 notes were played at those particular volumes; some frequencies will have been harmonics of lower ones. Fortunately, with a knowledge of how the instruments in question work, it's possible to predict where those harmonics should lie and how loud they would be. With a mixture of musical and scientific detective work, Brown was able to make a best guess as to the notes being played on Paul McCartney's bass, George Harrison and John Lennon's guitars and producer George Martin's piano. But there are other ways those same frequencies might have been produced, and it's fair to say that Brown's analysis didn't convince everyone. So it's likely we'll never know the ultimate answer to the mystery of the opening chord on *A Hard Day's Night*.

Harmonics aren't the only 'unplayed' frequencies that can occur in music. If you have two sine waves that are close together in frequency, then the combined waveform fluctuates up and down at a rate that is the difference between those frequencies. This is most noticeable when the difference in question is less than a semitone – which means, in a musical context, that someone is singing or playing out of tune. The effect is of a rapid pulsation – or 'beating', to use the technical term – of the combined sound, which slows down and eventually becomes imperceptible as the two tones are brought into sync. If you've ever wondered how musicians used to tune their instruments before electronic tuners were invented, that's the answer – they listened for the beating sound to disappear.

Now that I've explained the beating effect between notes that are slightly out of tune, I can go back and clarify something I touched on earlier in this chapter. You may remember I said that when Pythagoras tried to fit a twelve-note scale into an octave, it ended up slightly wonky. Today, we simply divide the octave into twelve equal semitones, but Pythagoras took a different approach. He wanted every note in the scale to be *exactly* a fifth higher than one of the other notes (allowing for possible transposition of a note to a higher or lower octave). The modern system gets close to that because every note is seven semitones higher than one of the other notes, and seven semitones is pretty close to the frequency ratio of 3:2 that Pythagoras wanted. But 'pretty close' wasn't good enough for Pythagoras – he wanted the relationship to be exact.

Unfortunately, Pythagoras was trying to achieve the impossible. To explain why, I need to go back to something else I said earlier: that the logarithm to the base two of 1.5 is 7/12 'as near as makes no difference'. But for Pythagoras to make the perfect twelve-note scale he wanted, it would have to be *exactly* 7/12, to the very last decimal place. And it isn't. So he ended up with a scale where eleven out of twelve notes met his requirement, but the last one didn't. It's sufficiently 'out of tune' that you can hear perceptible beating between it and the harmonics of the note that's supposed to be a fifth lower. The result sounds a bit like a wolf howling, and it's called a 'wolf fifth' for that reason. Even so, Pythagoras' wonky scale persisted well into the Middle Ages, before musicians took matters into their own hands and started looking for alternatives, finally settling on the system we use today.*

While we're on the subject of the frequency domain, there's one very basic point that may have occurred to you already. Taken literally, 'frequency' means how often something occurs in time – and with really low frequencies, like the beating of a wolf fifth, that's exactly how it sounds. On the other hand, we perceive higher frequencies – the so-called 'audible' ones above about 20 Hz, or 20 cycles per second – in a completely different way, as musical pitches. But that's just a reflection of the way our ears and brains work; from a mathematical and scientific point of view, there's no difference at all. In reality, an audible frequency is exactly what it says on the tin: a sound that repeats itself a given number of times per second.

---

* You can find plenty of videos on YouTube that illustrate the different tuning systems, for example: 'G Sharp and A Flat Are Not The Same Note', Jesse Strickland, https://www.youtube.com/watch?v=tGEXJe3px68

This leads to an intrinsic coupling between two properties of music, pitch and tempo, that we normally think of as completely different. If you take an analogue recording and play it back at twice the original speed, then all the frequencies will double too – in other words, it will sound an octave higher. This was well known in the days when vinyl records were ubiquitous, and most turntables could rotate at either 33 or 45 revolutions per minute. Choosing the wrong setting meant the song played not only at the wrong tempo, but the wrong pitch too. These days, however, digital processing means you can alter playback speed and pitch independently of one another.

In the pre-digital days, the pitch/speed coupling could cause headaches for music producers if they wanted to merge different takes of a song recorded at slightly different tempi. The most famous example of this is The Beatles 1967 masterpiece 'Strawberry Fields Forever'. This pushed the music technology of its time in many ways, but the particular aspect we're talking about here was more or less self-inflicted by John Lennon. His preferred sound was something like the first minute of Take 7 followed by the remainder of Take 26, but rather than simply re-record the whole song that way, he left it to producer George Martin to splice the two tapes together.

The problem was that Take 26 was recorded at 111 beats per minute (bpm), while Take 7 was quite a lot slower, at 90 beats per minute. In those days, when you couldn't alter the speed of a recording without changing its pitch, that might have made Martin's task impossible. But there was a saving grace: Take 7 was also recorded at a lower pitch, around 2.5 semitones below Take 26. So by speeding Take 7 up by 1 per cent and slowing Take 26 down by 11.5 per cent,

Martin was able to bring the two recordings virtually into sync – certainly to the point where Lennon himself couldn't spot the join.* Even so, there's a slight change in beats per minute from 91 to 98.

It's more than a little scary to think that, if you or I had the original takes, we could match them up perfectly – in both pitch and tempo – in just a few minutes using free software, such as Audacity or Waveform Free. But of course, if we did that, the result wouldn't sound anything like as timelessly brilliant as George Martin's version.

Another thing we've learned from looking at the frequency domain is that sounds differ, not only in their fundamental frequency, but also in the arrangement of higher frequency partials. Traditional musical instruments like strings and woodwind generally produce an orderly set of harmonics that are multiples of the fundamental, while more raucous sounds – like a vacuum cleaner, for example – produce a whole spectrum of partials across all frequencies.

It's tempting to think we've just hit on the perfect definition of music, as consisting primarily of harmonic partials, while everything else is mere 'noise'. But it's not that simple. A standard drum kit doesn't produce neat harmonics, and few people would dismiss a virtuoso drum solo as 'not music'. And as for vacuum cleaners – there are parts for three of them in Malcolm Arnold's *A Grand, Grand Overture* (1956), along with an electric floor polisher, four rifles and a standard concert orchestra. It may not be great music, but it *is* music.

A more serious musical work from the same decade is Edgard Varèse's 'Poème Électronique' of 1958, which

---

* It happens around the 60-second mark. For a more in-depth discussion, see 'Why is Strawberry Fields Forever in A half-sharp major?' by David Bennett: https://www.youtube.com/watch?v=QgtzOafdoOQ

incorporates various mechanical and animal noises as well as sirens, percussion and purely electronic sounds. As musicians go, Varèse may not be a household name, but he has a close link with one who is. Around the time of 'Poème Électronique', his work came to the attention of a teenage Frank Zappa – and made such an impression on him that he decided to switch hobbies from chemistry to music. And it was Varèse who came up with what, to me, is the ultimate definition of music: 'organised sound'.

To understand what that means, and its implications for the science of music, it's helpful to approach the subject in the way that a computer programmer would, in terms of parameters and algorithms. That's what we'll do in the next chapter.

MUSICAL ALGORITHMS 3

The *Oxford Dictionary* defines the word parameter as 'a measurable or quantifiable characteristic of a system'. So, since music is a system, we can reasonably talk about the parameters of music. After all, this is a book about the science of music, and scientists are the main users of the word parameter. Musicians may object that it's too cold and objective a word to apply to an art form, but on occasion it really has been used in that context, for example, by the 20th-century experimental composer Karlheinz Stockhausen. Despite his use of this and other snippets of scientific jargon, Stockhausen was very much a musician and not a scientist. As a young child, he was one of those annoyingly precocious prodigies who can hear a song on the radio and immediately sit at the piano and play it, chords and all. His likeness found its way into the crowd on the cover of The Beatles' *Sgt Pepper* album, and he was namechecked – along with Edgard Varèse, Igor Stravinsky, Arnold Schoenberg and many others – in the liner notes of *Freak Out!*, the first album by Frank Zappa and the Mothers of Invention.

In the case of music, quantifiable parameters include things like pitch, volume and note duration, all of which can easily be converted into numbers. But there's more to it than that. As with any kind of scientific analysis, the interesting thing isn't the numbers themselves, but the organising patterns and logical rules that connect them together. We can introduce another bit of modern jargon here and refer to these rules as 'algorithms'. Actually, that word isn't quite as modern as you might think – it was used, in the earlier form of 'augrim', by the 14th-century poet Geoffrey Chaucer, referring to the rules of arithmetic. And like 'parameter', it's not just science nerds like me who apply it to music. The musicologist Paul Griffiths – writing in 1995, before anyone had heard of search engine algorithms or social media algorithms – talked about 'the importance of rules and algorithms in composition'. That's going to be the main subject of this chapter, but to start with, let's take a closer look at the raw material of musical algorithms: the parameters.

Parameters of music

One of the most basic elements of music is the rhythm, or the arrangement of notes in time. At the very simplest level, this is set by just two parameters: the time at which a note starts playing, technically called its onset, and its duration. If you were to look inside a MIDI file, which tells a digital instrument how to play a piece of music, then all the notes really are characterised by those two numbers. But if you look closer, there's another important parameter too – and that's the volume of a note.* Even within a single bar of music, some notes are meant to be stressed – in other words, played more loudly than others. This pattern of stresses reflects the metre of the music, which is what people really mean when they talk about its rhythm.

In music, metre works much the same way as in poetry. Take a line of poetry with a clearly defined metre: 'Once upon a midnight dreary, while I pondered, weak and weary' from Edgar Allan Poe's *The Raven*, for example. That has sixteen syllables, which you could set to music as four bars with four beats per bar. A 'bar' is the basic repeating unit of the rhythm, and the beats are equal subdivisions of a bar. They're played with different stresses, roughly as you'd read Poe's poem: strong-weak-medium-weak. The result is the commonest of all musical metres, reassuringly referred to as 'common time'. The length of a beat is a quarter of a bar, and a note of that duration is known, perfectly logically, as a quarter note. The metre is notated with a 'time signature' of 4/4, indicating four quarter notes in a bar, and notes of other lengths are named accordingly, with half a quarter note being an eighth note and so on.

Looking rather like a mathematical fraction, the time signature is a conventional way to indicate the metre of a piece of music, or of a section within a larger piece. The first or upper number indicates the number of beats in a bar, while the second or lower number specifies the length of a beat. Time signatures aren't found in all types of music because not all music has a clearly defined metre. Exceptions

---

* Actually, the technical word is 'velocity', but there's so much scope for misunderstanding that term, particularly if you're unfamiliar with music software, that I'm going to use everyday words like volume and loudness instead.

include medieval plainsong and other chant-based religious music, or the improvisatory cadenzas in classical concertos. The more improvisatory-sounding parts of some progressive rock songs, such as Pink Floyd's 'Interstellar Overdrive' and King Crimson's '21st Century Schizoid Man', also lack time signatures, as do some more recent experimental electronic tracks, such as 'Mercy Funk' by Venetian Snares. But such examples are very much the exception rather than the rule.

Not all time signatures are as thoroughly logical as 4/4. If you have a song with three beats in a bar, such as the verse section of 'Lucy in the Sky with Diamonds' ('**dum**-dah-dah **dum**-dah-dah **dum**-dah-dah **dum**-dah'), then the time signature is written as 3/4 – i.e. three quarter notes per bar, even though they're not literally a quarter of anything any more. This particular metre, 3/4, is often called waltz time, after the main dance style that uses it. But it was also used in several older dances such as the minuet and polonaise – but not, as mathematicians might imagine, in the tarantella, which is notated with a time signature of 6/8. Despite the fact that 'six-eighths' sounds like it ought to equal 'three-quarters', it's a different rhythm altogether.

If you play 3/4 in eighth notes – i.e. three beats, each of which is split into two – then it comes out as 'one-and-two-and-three-and'. On the other hand, 6/8 is more like two beats, each split into three: 'one-and-a-two-and-a'. If that sounds confusing, just think of the song 'America' from *West Side Story*, which has bars of 6/8 ('dadadah, dadadah') alternating with 3/4 ('dah, dah, dah').

Besides 4/4, 3/4 and 6/8, there are a couple of other time signatures that crop up quite often, such as 2/4 and 12/8, but that's about it as far as most genres are concerned. Occasionally, for dramatic effect, composers will effectively run 2/4 and 3/4 together to get 5/4, as in the 'Mars' section of Gustav Holst's *Planet Suite*, or Lalo Schifrin's original version of the *Mission Impossible* theme. But when Damon Albarn used a 5/4 metre in the second song on Gorillaz' eponymous debut album, it was still enough of an oddity that '5/4' ended up being the song's title.

The more unusual the time signature, the more disconcerting it sounds to our ears – in many cases precisely *because* it's unusual. What may be the most awesomely effective use of 'uncommon time' of all came about as the result of an accident. I'm referring to the electronic title music that Brad Fiedel wrote for the *Terminator* movie in 1984. It was supposed to be in 6/8, but something went slightly wrong when Fiedel was programming his sequencer and it ended up in 13/16 instead. The result, with its unsettlingly irregular beat, is guaranteed to send chills down your spine.

I was careful to say the range of time signatures is limited in 'most genres'. The fact is, there are a handful of specialist genres that positively revel in weird metres. Many of these have 'math' in the name, like math rock, math metal and mathcore. Personally, I'm not convinced that counting higher than four really counts as 'mathematics' – I mean, I wouldn't put it in the same league as logarithms and set theory (which we'll encounter later in this chapter) – but they're still great genre names. That can't be said for what may be the most important genre specialising in irregular rhythms: the confusingly named 'intelligent dance music', or IDM.

The genre seems to have arisen in the 1990s as an off-shoot of electronic dance music (EDM). When people like Aphex Twin (Richard D. James) and Autechre (Rob Brown and Sean Booth) started experimenting with odd time signatures and irregular rhythms in their music, it got to the point where it was no longer danceable. It was still electronic music, but it was music for listening to or thinking about rather than dancing to. Logically it should have been called 'electronic listening music' or 'electronic thinking music'. But the marketing people went and changed the wrong word, so we're stuck with 'intelligent dance music' – a term that's unanimously hated by everyone in the field. Its name aside, though, IDM is one of the most interesting and innovative genres around today, and we'll come back to it in a later chapter.

Another way to introduce odd rhythms into music is to have different instruments playing in different metres, producing so-called 'cross-rhythms'. There are two possible approaches here. The first, which can be found in some math rock songs, is for everyone to play at the same bpm, but with a different pattern of strong and weak beats – so that effectively they're playing bars of different length. '5/4' by Gorillaz is an example of this because the drummer doggedly insists on playing in standard 4/4 time – with bars four-fifths as long as those of the main song in 5/4.

Going back a couple of centuries, Mozart took the idea even further in the first act finale of his opera *Don Giovanni*, which sees three different dances going on at once. To start with, there's a stately minuet in 3/4 time, but then the title character embarks on a country dance in 2/4. It's at the same bpm, so Don Giovanni's bars are only two-thirds the length of the minuet's bars. Finally, a fast peasant dance starts up, which is notated in 3/8 so it gets through two bars for every one bar of minuet. This might sound like a recipe for disaster, but Mozart (being Mozart) fits it all together very neatly.

The other way to combine metres is to have everyone playing bars of the same length, but with a different bpm to fit a different number of beats into the bar. So, for example, one instrument may play two beats in a bar while another plays three beats in the same bar. The technical name for this is a hemiola, and it was a favourite motif of the composer Brahms in the 19th century. Hemiola-like cross rhythms are also a fundamental characteristic of much of the music of sub-Saharan Africa.

As cross-rhythms go, the hemiola is a quite a mild one. A musician who took the idea much further was Frank Zappa, who coined the term 'rhythmic dissonance' for this kind of thing. Just as harmonic dissonance is a clash between ill-matched frequency ratios, what we have here is a clash between ill-matched time signatures. Around halfway through the avant garde piece 'Toads of the Short Forest' on his 1970 album *Weasels Ripped My Flesh*, Zappa is heard to say: 'At this very moment on stage, we have drummer A playing in 7/8, drummer B playing in 3/4, the bass playing in 3/4, the organ playing in 5/8, the tambourine playing in 3/4 and the alto sax blowing his nose.'

Besides rhythm, the other fundamental element of music is the 'tune' – or melody, to use a slightly more technical term. In principle, the notes of a melody could be characterised with just a single additional parameter, pitch, on top of the onset, duration and volume that we've already introduced. In fact, that's exactly what a digital MIDI file does – it uses, for example, the numbers 21 to 108 to represent the 88 pitches of a standard piano keyboard. For us humans, however, this could get a bit unwieldy, so it's easier to break it down into two separate parameters. One of these simply specifies which octave the note falls in, while the second, more important number indicates whereabouts in the octave – on which of the twelve semitones – the note falls.

Returning to the terminology used in the previous chapter, we can refer to the twelve notes of an octave as 'pitch classes'. This concept actually comes from the mathematics of set theory (I warned you it was coming up). There's nothing complicated about the concept of a set; it's just a collection of items that the user intends to work with. In the case of musical PCs, these constitute a set of twelve numbers, from zero to eleven, with no higher or lower values. If you add, say, five to nine, you can't get fourteen because it doesn't exist in this system. You need to subtract twelve to get the real answer, which is two. I could describe this as 'modulo-twelve' arithmetic, but that would just confuse things further (apart from for sci-tech nerds). It's easier to think of PCs like the numbers on a clock face, which is almost the same except that the clock runs from one to twelve instead of zero to eleven. And everyone knows that adding five hours to nine o'clock gives you two o'clock.

At the risk of annoying a few readers, I'm going to say that – *for the great majority of listeners and musicians* – the absolute values of the PCs in a melody are almost completely irrelevant to its musical effectiveness. The thing of crucial importance isn't the pitch of a note, but the 'interval', or difference in pitch, between one note and the next. So, alongside PCs, we also need to talk about interval classes (ICs).

Musical set theory defines six different ICs, covering all the possible values of the minimum distance, measured in semitones, between one PC and the next. The shortest distance is one semitone – in this context, repeated PCs are treated as the same note, so we don't count zero – and the largest distance is six semitones, or half an octave. That's because when we look for the minimum distance between two notes, we consider both directions. Going from

C to A may be nine semitones, but going from A to C is just three.

To most people, one PC sounds very much like another if it's heard in isolation outside any particular musical context. But when it comes to ICs, the situation is completely different; they all have very distinctive sounds. Some of them we've met before – for example IC 1 is the 'wrong note' sound of the semitone, while IC 6 is the tritone 'Devil's chord'. IC 2 is a whole tone – the step between the notes 'do-re-mi' when you sing a simple scale. IC 3 and 4 are the minor and major thirds, which have very different sounds despite their similar names.

As for IC 5, it would be neat to think it's the 'fifth' we met in the previous chapter and in fact it is, although that's not how it got its name. As we've already learned, a fifth is seven semitones – but we're in a modulo-twelve world here and going up seven is the same as going down five. That's where the 5 in IC 5 comes from, and musically (just to confuse things a little more) five semitones is called a fourth. What we've just shown is that mathematically, a fourth is identical to a fifth, because they both represent IC 5 – one going up, the other down. A musician would say that a fourth is the 'inversion' of a fifth.

Earlier in the book, I promised I'd eventually get round to explaining where musical terms like 'third', 'fourth' and 'fifth' come from. We've almost got to that point now. The thing is, the division of the octave into twelve equally spaced PCs is great for scientific analysis, and – despite the impression I may have just given – it really does make mathematical calculations much simpler. That means it's great for computers and MIDI keyboards too. But it isn't great for musicians because it turns out to be almost impossible to write comprehensible music that uses all twelve PCs without imposing any other constraints. In the first couple of decades of the 20th century, a few composers such as Arnold Schoenberg tried to do that, and quickly discovered that there was such a thing as having too much freedom.

After a lot of experimenting, Schoenberg came up with a clever system that allowed him to use all twelve PCs – essentially by putting constraints, not on the PCs themselves, but on the allowable intervals between them. It's actually quite a 'scientific' system – and a fascinating (if frequently misunderstood) one – so we'll take a more detailed look at it later in this chapter. For the moment, though, we'll focus on a more traditional approach based on specifying the range of permissible PCs rather than ICs.

Looking at a piano or MIDI keyboard, you're struck with the immediate impression that not all PCs are equal. There are white keys and black keys, and the white ones are larger and more accessible than the black ones. If you look for the repeating pattern of keys, you'll see that of the twelve notes in an octave, seven are white and five are black.

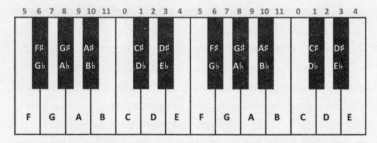

Two octaves of a piano keyboard, with notes labelled with their traditional names – of which there are two possibilities in the case of the black keys – and the corresponding pitch class numbers along the top.

It's reasonable to deduce that there's something special about the seven white notes.

In the earliest days of European music, long before keyboards were invented, only the seven 'white notes' were used. Variety was added by starting and ending melodies on different notes, which gave the seven medieval 'modes'. For example, the Latin hymn 'Ave Maris Stella' starts and ends on the note D, which puts it in the Dorian mode. This is one of several modes that fell out of favour in Renaissance times but has enjoyed a resurgence since the middle of the 20th century – first in jazz, then in all kinds of popular music.

That means the Dorian mode, which someone like Mozart might have dismissed as hopelessly old-fashioned, can be heard in, for example, 'So What' by Miles Davis, 'Another Brick in the Wall' by Pink Floyd and 'Uptown Funk' by Bruno Mars. Another relic of the Middle Ages, the impressively named Mixolydian mode is encountered even more frequently today. It's the white-note scale starting on G, and features in numerous popular hits including 'Paperback Writer' by The Beatles, 'Man in the Mirror' by Michael Jackson and 'Shake It Off' by Taylor Swift.

But for three centuries prior to the Jazz Age, the Dorian and Mixolydian modes – and the ones starting on E, F and B as well – were virtually ignored at the expense of just two modes: the ones starting on C and A. They had different names in the past, but today those are called the 'major' and 'natural minor' modes respectively. The major scale is the one that goes 'do-re-mi-fa-so-la-ti-do', where the two notes ending in '-i' are followed by a semitone and all the other notes by a whole tone. Its most distinctive characteristic is that the third note, E, is four semitones (IC 4) above the starting note C. As we learned earlier in this chapter, IC 4 is what musicians call a major third – a revelation that may well have you shouting 'Aha!' at his point. Because the mystery is finally solved – it's called a major third because it's the *third* note of a *major* scale. And if you carry on counting, you'll see that five semitones – a musical fourth – gives you the fourth note, and seven – a fifth – gives you the fifth note. So there is a kind of logic to those names after all.

The fourth and fifth don't need a 'major' qualifier, because if you look back at the keyboard and count up semitones starting on A, you'll see that five and seven take you to the fourth and fifth notes of that scale too. But the third is now just three semitones, or IC 3 – which, reassuringly enough, we defined earlier as a 'minor third'.

There are limits to what you can do with just the white-note major and minor scales, both in terms of musical expressiveness – it's difficult to convey change, motion or tension when you only have seven PCs to play with – and for purely practical reasons. It may be that a song would be easier to sing, or would sound more natural to the ear, if it was, say, a fifth lower. But you can't drop a whole melody by a fifth and stay on the white notes. This is where the black notes finally come into their own, because they allow music to be transposed by any number of semitones up or down, and still preserve the exact shape of the melody.

Technically, when you transpose music up or down, you're putting it in another key. The white-note major key is C major, but if you transpose up by three semitones you get Eb major; similarly, going two semitones down from A minor takes you to G minor. For many years in the 18th and 19th centuries, keys were considered so important by classical composers that they often named works after them – with their more helpful modern nicknames being added at a later date. So, for example, Haydn's 'Farewell Symphony' was the 'Symphony in F♯ minor', Schubert's 'Death and the Maiden Quartet' was his 'String Quartet in D minor' and Chopin's 'Minute Waltz' was 'Waltz in D♭ major'. This bizarre practice has always baffled me, but I think it may have originated in the slightly different way that listeners perceived music in those days – a subject I'll come back to in the final chapter.

As widespread as major and minor keys are in Western music, there's nothing really fundamentally significant about them. When it comes down to it, they're just a cultural choice. You could argue, as Pythagoras did, that the twelve-PC scale *does* have a fundamental significance, because, with a little fudging, it meets his criterion of every PC being a fifth higher than one of the other PCs. But as soon as you narrow things down to a subset of those twelve notes, you're doing something that's basically artificial. You can see that if you look at the music of other cultures around the world because they all divide the octave in different ways.

To pick just one example with a particularly satisfying logic behind it, we'll take a brief look at Japanese scales. It's worth recalling that the medieval European modes used a fixed set of seven notes – the white keys of a piano – and then chose any of those notes to start and end a scale on. The Japanese system is similar, but with two important differences. There are only five notes, not seven, and they're not always the same five notes. But they can't be any five notes, they have to be chosen carefully.

If you remember, we said that in Western music, both the major and minor scales share the same fourth (five semitones) and fifth (seven semitones) above the starting note. In the case of C major, the fourth is F and the fifth is

G. Those three notes are structurally very important to all types of music, so we'll keep them – which means we've got three-fifths of our five-note scale already. If you look back at the keyboard diagram, you'll see the remaining notes form two similar patterns, one running from D♭ to E between C and F, and the second from A♭ to B between G and the upper C. It seems a shame to break this symmetry, so we'll choose the corresponding notes from each pattern as our final two notes. In other words, either D♭/A♭, or D/A, or E♭/B♭, or E/B.

All the resulting scales, and the various modal permutations of them, are valid in the context of Japanese music. And to ears accustomed to Western music, some of them, particularly the first, have a very 'exotic' sound. So, if you've ever wanted to write an opera about, say, the Emperor of Japan banning extra-connubial canoodling, now you know how to do it.*

Music's third dimension

The elements of music that we've talked about so far, rhythm and melody, can be visualised as occupying a two-dimensional space. The rhythmic parameters – note onset and duration – determine how the music plays out in time, which is usually portrayed as a horizontal axis running from left to right. On the other hand, pitch, the main melodic parameter, can be thought of as running from high to low on a vertical axis. If music can be said to have a third dimension, then it's

---

* Yes, I know it's already been done, see: 'No. 5 Entrance of Mikado & Katisha', Gilbert and Sullivan Archive, https://www.gsarchive. net/mikado/webopera/mk205.html

provided by what musicians refer to as 'harmony' – the effect created by playing multiple notes simultaneously.

To the ear, the addition of harmony really does give music a sense of 'depth' that is quite distinct from the horizontal and vertical dimensions of rhythm and melody. But from a purely objective point of view, it can still be fitted into a two-dimensional picture. You'll be aware of that already if you can read sheet music, or if you've ever played with the 'piano roll editor' in GarageBand or similar software. Here's an example from the dramatic opening of Brahms' Symphony No. 1:

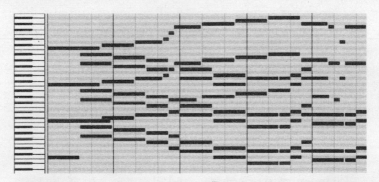

A 'piano roll' representation of the opening of Brahms' first symphony, showing the complex vertical harmonies in addition to the horizontal motion of the music.

The keyboard on the left-hand side shows which notes are being played (running from low pitches at the bottom to high ones at the top), while the horizontal dimension indicates when they're played, relative to the vertical bar lines. It's clear that harmony – the cumulative effect of many different notes heard simultaneously – plays a crucial role here. The music would lose all its threatening power – in fact, it would be pretty meaningless – if all you could hear was a single melodic line.

In the case of melody and rhythm, we saw how it's possible, through the use of time signatures and scales respectively, to organise music in a way that sounds ordered and logical rather than totally random. The rules in those cases scarcely warrant the name 'algorithms'; they're simple enough that children can pick them up instinctively without any conscious effort. But it's a different matter when we come to harmony, because the potential parameter space is so huge. If we don't do some serious rulemaking, it's all going to end in a random cacophony.

The notes shown in the Brahms example span four octaves, but as far as the basic harmony is concerned, we can ignore a note's octave and just focus on its pitch class. That still provides a large number of note combinations, or 'chords', but most genres are dominated by a small subset of these. In fact, just two chord types – or perhaps only one – account for most of the chords you'll hear in everyday music. These are called major and minor triads, and they're related to the major and minor scales we met earlier.

Starting with the major triad, this consists of the first, third and fifth notes of the major scale. So, for example, C major is made up of C, E and G, or PCs 0, 4 and 7. If you work out the corresponding interval classes – the shortest distance between any pair of notes in the chord – you end up with IC 4, 3 and 5. As for the minor triad – as you might guess, that's made up of the first, third and fifth notes of the *minor* scale. In the case of C minor, that would be PCs 0, 3 and 7 – which, as you'll already know if you're a music lover, has a completely different sound to a major triad. Yet the ICs, 3, 4 and 5, are exactly the same. Mathematically, there is hardly any difference between a major and minor chord; one is simply the inversion of the other. Start on any note, go up four semitones, then up another three and you have a major triad. Start on the same note, go down four semitones, then down another three and you have a minor triad.

The astonishing thing is that, between those two, mathematically almost identical, chords, you have the basis for the bulk of all the harmonies to be found in most types of music. It's not uncommon to add other notes to the three basic ones to add dramatic or emotional effect. For example, adding a B♭ to the C major chord – to give 'C7' in music jargon – introduces the more bluesy-sounding intervals IC 2 and 6. But even with four-note chords like this it's the embedded three-note triad that defines the basic harmony.

Since there are twelve PCs, and you can build either a major or minor triad starting on any of them, there are 24 triads altogether. That's still quite a lot of chords to navigate our way through, and it's useful to have some kind of map of them. A handy diagram in this context is the 'Tonnetz', which comes from the German for 'tone network'. The nodes in this network (of which the picture overleaf is just a part) correspond to PCs, arranged along three intersecting sets of lines. Along the horizontal lines, the PC number goes up by a fifth at each step. From top left to bottom right it goes up by a major third, and from bottom left to top right by a minor third.

Each of the small triangles is a triad chord; upward pointing triangles are minor triads, downward pointing ones are major triads. Every chord of one type has three chords of the other type adjacent to it, each of them differing by just one PC. This allows us to navigate from any given triad to any other by a series of transformations.* As an example, let's take the E♭ major chord. If we flip the triangle upwards, we end up with E♭ minor, while moving to the left takes us to C minor and to the right goes to G minor. These three manoeuvres are referred to as 'P', 'R' and 'L' transformations, respectively (the letters standing for music-theory terms we don't need to go into here). All three transformations work equally well in both directions, either from major to minor or vice versa.

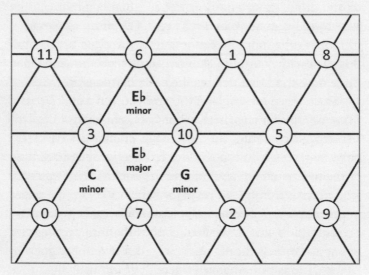

A 'Tonnetz' diagram illustrating relationships between chords. The circled numbers are PCs and the triangles are chords, with a few labelled examples.

---

* Mathematicians will realise that we've just graduated from set theory to group theory, but fortunately, we don't have to go into any detail about that here.

So now, by concatenating P, R and L transformations, we can jump from any chord we like to any other. I'll try to give a simple example to show how this works. It's not going to be as easy on the printed page as it would be in a video or podcast, but hopefully if I refer to the 'The Imperial March' from *The Empire Strikes Back* – '**dah, dah, dah,** *dah-di* **dah,** *dah-di* **dah**' – then most people will be able to hear it in their head. The chords corresponding to bold type, as I've written it, are in G minor, while those in italics are in E♭ minor. Looking back at the Tonnetz diagram reveals that this sinister-sounding chord shift is an 'LP' transformation – first left (to E♭ major, which we leapfrog over), then up. It's as simple as that.

In the jargon of music theory, the LP transformation is referred to as a 'chromatic mediant'. It's a common, almost clichéd, occurrence in certain types of film music, precisely because of its sinister, unsettling effect. But you'd be hard pressed to find it anywhere in the classical music of the 18th or early 19th century. In those days, composers had strict rules regarding 'allowable' chord progressions, and the chromatic mediant simply wasn't on the list. That it eventually gained its freedom was largely down to Richard Wagner – another composer who, along with Franz Liszt, focused on creating the *sound* he wanted, rather than following some stuffy academic's idea of the rules. As a result, the opening of the third act of Wagner's opera *Siegfried*, which is full of chromatic mediants, sounds almost like a Hollywood movie score to modern ears.

If Liszt and Wagner liberated some previously forbidden triadic chord progressions, they also made the first tentative steps towards a much bigger musical revolution – freeing up all those non-triadic harmonies. There are so many possibilities here that they weren't fully enumerated until

1973, when the music theorist Allan Forte did just that in a book called *The Structure of Atonal Music*. It may be worth a brief aside at this point to look at that odd word 'atonal'. In English, it sounds like it means 'lacking tones', which suggests something like a drum solo. But that's not what atonal means at all. It helps to know that the French word for 'key', in the musical sense, is *tonalité* – so, for example, when Liszt called one of his piano pieces 'Bagatelle sans tonalité' it simply means 'bagatelle without a key'.

And that's what Forte's book is about – not music without tones, but music without a specific key. Or to put it another way, music that doesn't stick to the notes of a particular major or minor scale. That frees up every possible combination of three or four PCs to serve as a chord. If you start each combination on the note C (PC 0), then Forte counted nineteen three-note chords and 43 four-note ones. In most cases (but not all, as we'll see in a moment), you can multiply these numbers by twelve because you could start the chord on any PC you like. That comes to a huge number of chords – and while some of them sound truly awful, there are others that have been unjustly neglected by mainstream music.

Once again, it's useful to have a visual representation of all these different chords. The fact that they're constructed from 'modulo-twelve' PCs – which, as we learned earlier, work a bit like the numbers on a clock face – gives a strong hint as to how to go about this. We can use musical 'circle diagrams' to depict chords as polygons inside a clock-like dial running from PC 0 to 12.

The first chord shown here is the major triad. The most noteworthy thing about it is just how 'unspecial' it looks, for a chord with the weight of centuries of music theory on

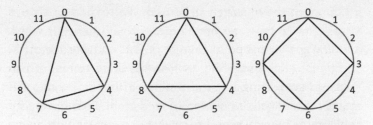

'Circle diagrams' illustrating, from left to right, a major triad, an augmented triad and a diminished seventh chord.

its shoulders. I haven't shown the minor third, which has a three in place of the major triad's four, because it's really just an upside-down version of the same shape. In contrast, the chord in the middle looks much more striking. It goes by the technical name of an 'augmented triad', and geometrically it's a perfectly symmetrical equilateral triangle. All the PCs are separated from one another by a major third or its inversion.

The symmetry of the augmented triad gives it another unique property. I said a moment ago that there are twelve versions of 'most' chords, because you can start them on any PC you like. But in the case of the augmented triad, there are only four distinct versions, because rotating the triangle by four semitones simply maps it onto itself. So you can have augmented triads starting on PC 0, 1, 2 and 3 (i.e. C, C♯, D and E♭), but after that you're just repeating yourself.

Personally, I love the augmented triad, as much because of its fascinating symmetry as for its rather spooky sound. Yet of all the three-note chords with well-established names, it's by far the least used. That's because it's fundamentally 'atonal', in the sense of not belonging to any key – and being almost-but-not-quite a conventional triad in no fewer than six different keys.* When it is used, it's usually as a brief 'passing note' between one chord and another. But when it's allowed to stand on its own, it can have a truly unearthly effect – as in the segment of Iron Maiden's 'Rime of the Ancient Mariner' when, after a supernatural dice game, the souls of the Mariner's crewmates are forfeit: 'Four times fifty living men, / (and I heard nor sigh nor groan) / with heavy thump, a lifeless lump, / they dropped down one by one.'**

The final chord I've shown is another symmetrical one, the diminished seventh. Here the separation between each pair of adjacent PCs is a minor third, while that between opposite pairs is a tritone. In this case there are only three independent chords, starting on C, C♯ and D. This gives the same tonal ambiguity as the augmented triad, but – I guess – a sound that was considered more acceptable in the 18th and 19th centuries. For whatever reason, the diminished seventh had the opposite fate to the augmented triad, being used in all kinds of music to represent extremes of stress or emotion:

> Wherever something difficult was to be done, one enlisted the services of this miracle worker, equal to every task ... Wherever one wanted to express pain, excitement, anger, or some other strong feeling – there we find, almost exclusively, the diminished seventh chord.

Those words were originally written in 1911, in a 400-page textbook called *Theory of Harmony* by Arnold Schoenberg.

---

* It's just one semitone away from three major chords and three minor chords.
** These lines come from the 1817 edition of 'The Rime of the Ancient Mariner' by Samuel Taylor Coleridge, so they are safely in the public domain.

It's a fascinating book to browse through, with two points coming across very strongly: first, that Schoenberg had an impressively complete knowledge of classical music theory, and second, that he loathed the subject with a vengeance.* Or at any rate, he hated its clichés and entrenched conventions – such as the over-use of the diminished seventh just referred to – and he wasn't afraid to say so. As the editor says in the introduction to the version on the Internet Archive:

> Practically every topic, be it the major scale or the minor, be it parallel fifths and octaves, non-harmonic tones, modulation, or the minor subdominant, leads him into wide-ranging speculations on nature, art, and culture, and vigorous attacks upon ossified or irrational aesthetics.

What Schoenberg could see, along with a relatively small number of earlier composers and countless later ones, was that musical culture should never stand still. In order to survive, music must evolve.

Musical evolution

Like all forms of art, music changes over the years to suit changing fashions, but there's more to musical evolution than that. One obvious factor is the increasing sophistication of music technology – particularly with the advent of electronics in the 20th century – and that's a subject we'll look at in depth in the next chapter. But there's been another, more

---

* The book can be downloaded free of charge from the Internet Archive: https://archive.org/details/SchoenbergArnoldTheoryOfHarmony/

subtle trend too. For over 200 years, there's been a steady broadening-out of the very *language* of music, in terms of the permissible ways that rhythms, melodies and harmonies can be put together.

This expansion of musical language has been driven primarily by the desire for a greater range of expressiveness, especially when it came to conveying extremes of emotion that hadn't previously featured prominently in music. While music has always been profoundly emotional, in the 18th century (when Europe's conception of 'correct' musical practice first crystallised) composers tended to exercise more decorum in the way they displayed it. The 'passion' in Bach's *St Matthew Passion*, for example, is all about religious devotion, sadness and mourning, while Mozart's operas like *Don Giovanni* focus on the classic triad of love, longing and jealousy. In those contexts, you can stay safely within the bounds of musical 'niceness'. So, Mozart, who died in 1791, could write that 'music, even in the most terrible situations, must never offend the ear'.

By the next generation of composers, things had begun to change. This was the dawn of what historians call the 'Romantic' age – across all forms of art – but they're using the word to refer to wild and extreme passions of any type, not just amorous ones. The turning point is often taken to be Beethoven's 'Eroica Symphony' of 1805 – which, possibly for the first time in history, introduced deliberate ugliness into music. That's not meant as an insult, although it might have seemed that way to Mozart. There were plenty of ugly scenes in the novels of that time – as there had been in Shakespeare's plays and the ancient Greek tragedies – so why single out music for a taboo against ugliness? Whatever the reason, Beethoven broke it with 'Eroica'. The emotional climax of the first movement is a series of ear-splitting, staccato dissonances which Wikipedia describes as an 'outburst of rage'.

Another striking example, from 1827, is 'Frühlingstraum' ('Dream of Spring') from Franz Schubert's great song cycle *Winterreise*. If possible, try to find the version recorded by Peter Pears and Benjamin Britten, because they milk the effect I'm referring to for all it's worth. As you start to listen, you may wonder what on earth I'm talking about because the song begins with a beautiful melody that Mozart might have wished he'd written. But that's the 'dream' part. After about 40 seconds the narrator wakes up to harsh reality. Not only does the melody become much more jagged, but it's punctuated by a series of increasingly dissonant chords. In its time, the music must have sounded as shocking and outrageous as punk rock did to audiences in the 1970s. As Britten observed: 'I'm sure that one of the things that shocked them at that first sing-through of the piece was that he could write so beautiful a tune and then interrupt it so horribly with these famous chords.'*

Although Beethoven and Schubert weren't afraid to put ugly sounds into their music, they still employed the same basic building blocks, in terms of scales and chords, that Mozart had done. They simply used those building blocks in, for want of a better word, a 'noisier' way. The same trend has continued in much of Western music to the present day. While most rock music, for example, is louder and more in-your-face than a Schubert song, it's really not that different harmonically. If you don't believe me, have a listen to

* 'Peter Pears & Benjamin Britten discuss "Die Winterreise"', kadoguy, YouTube, https://www.youtube.com/watch?v=MFaH-Kb2HD0

my 'heavy rock' arrangement of Schubert's song 'Erlkönig' (https://www.youtube.com/watch?v=u3nvuNC19vk). In certain corners of classical music, however, even the ways in which melodies and chords are put together have evolved since the middle of the 19th century.

Two composers we've already encountered in this context are Liszt and Wagner. Both played important roles in the evolution of music, but in rather different ways. Of the two, Wagner was much more stridently vocal about the need for music to change. He even wrote a long essay in 1860 called 'Music of the Future' – by which, of course, he meant his own music. This was around the time of his opera *Tristan and Isolde*, which created a whole new vocabulary of yearningly romantic chords that transformed the way love scenes are depicted in music. And we've already seen how *Siegfried*, a decade later, introduced the clichéd sound of the chromatic mediant, which remains a Hollywood favourite to this day.

Despite this, Wagner wasn't the towering musical revolutionary he made himself out to be. It's true that he created his share of brand-new sounds – in the form of chords and chord progressions – but it was all done within the same tonal framework as Beethoven or Schubert's music. For a real revolutionary, we have to turn to Wagner's near contemporary Franz Liszt. He seems to have been the first person – in post-medieval Europe, anyway – to realise that there is no God-given reason why music has to rigidly stick to the 'tonality' of keys and seven-note scales.

I've already mentioned Liszt's 'Bagatelle sans tonalité' and, in the previous chapter, his Mephisto Waltz No. 2. Both pieces borrow so many PCs from outside a standard seven-note scale that it's difficult to say what key they're written in (and rather pointless to try). The opening of his

'Faust Symphony' (1857) contains all twelve PCs one after the other, arranged like a descending sequence of augmented triads: 7, 11, 3, 6, 10, 2, 5, 9, 1, 4, 8 and finally 0. You can't get further from 'tonality' than that.

The idea of accessing the full range of PCs, rather than a subset of them, seems to have appealed to Liszt. In a letter in 1860, he drew a chord containing all twelve PCs, which he said 'will thus form the unique basis of the method of harmony – all the other chords, in use or not, being unable to be employed except by the arbitrary curtailment of such and such an interval'. By the first decade of the 20th century, this suggestion – which Liszt may have meant as a joke – was starting to look distinctly prophetic. Many of the 'avant-garde' (i.e. consciously establishment-subverting) composers of that time were writing music that was 'atonal' in the sense of not being confined to a specific key. This could be said, for example, of Charles Ives in the United States, Claude Debussy in France and Alexander Scriabin in Russia.

But the composer most connected with the development of atonality is one we met earlier: Arnold Schoenberg of Austria. As we've already seen, he was thoroughly fed up with the conventions of classical music and wanted to get as far away from them as possible. While Ives, Debussy and Scriabin may have tweaked the rules, Schoenberg threw them away altogether. This was a stressful time in his life, both for personal reasons and due to the looming threat of all-out war in Europe. As a result, much of his earliest atonal music is screechingly angst-ridden and, in many cases, clearly designed to shock the audience (the 'punk rock' syndrome again). Yet this period did produce one genuine masterpiece: the song cycle *Pierrot Lunaire* of 1912. With its cabaret-influenced sound, it's not so much avant-garde as a precociously early example of 'avant-pop' (indeed, the best-known of all avant-pop artists, Björk, has performed *Pierrot Lunaire* in concert).

So, Schoenberg started his musical revolution by throwing away the rule book. Before long, however – once he'd worked off his anger and frustration – he realised that music *needs* some kind of rule book. To repeat a quote I mentioned earlier, from Schoenberg's French–American contemporary Edgard Varèse, music is 'organised sound' – and rule-following is the most obvious way to organise it. But Schoenberg didn't go back to the old rules, he came up with a whole new set of his own, called the twelve-tone method.

It's worth going into a little detail here, because (bearing in mind the title of this book) Schoenberg's twelve-tone system is considerably more 'scientific' than traditional music theory. Sadly, for that very reason it's often derided by musicians and listeners alike, who feel there's no place for science or mathematics in music. Presumably, they've never actually bothered to listen to a range of twelve-tone music, because then they'd realise that it can't be dismissed with a single sweeping adjective like 'cold' or 'impersonal' or 'artificial'. I'm sure some twelve-tone music falls into those categories, but Alban Berg's beautiful Violin Concerto, or Schoenberg's own harrowingly emotional 'A Survivor from Warsaw', certainly don't.

The point is, the twelve-tone system is a *method of composition*, not a genre. It's a method with an enormous amount of freedom, which can be used to create relatively 'mainstream' music as well as more challenging avant-garde works. It's been used in movie music, most notably in Jerry Goldsmith's score for the original 1968 *Planet of the Apes*, but also in a number of British horror films of a similar vintage, including *Curse of the Werewolf* (1961) and *Dr Terror's House of Horrors* (1965). Schoenberg's arch-rival Stravinsky used it in his *Elegy for JFK* (1964) and, going from the sublime to the ridiculous, his setting of Edward Lear's nonsense poem 'The Owl and the Pussy-Cat' (1966). As a teenager, even Frank Zappa turned out a twelve-tone piece called 'Waltz for Guitar', which ended up as part of the Mothers of Invention song 'Brown Shoes Don't Make It' (1967). My own contributions to the twelve-tone canon, which you can find on my YouTube channel, fall in the disparate genres of lo-fi hip hop and heavy metal.*

Despite its name, the twelve-tone method doesn't revolve around specific 'tones' – i.e. pitch classes – so much as around the intervals between them. In fact, it's tempting, given that Schoenberg was born in the same decade as Einstein, to think of it as a musical 'theory of relativity'. That's not an original idea; the composer Pierre Boulez wrote that 'with the 12-tone system, music moved out of the world of Newton and into the world of Einstein'.

As it happens, Einstein and Schoenberg really did have a loose acquaintanceship. They shared a profound concern for the plight of Europe's Jews, leading to a series of letters and at least one meeting between the two in the 1920s and 1930s. But there's no way the work of either one influenced the other. When questioned on the matter in 1933, Schoenberg replied: 'There may be a relationship in the two fields of endeavour, but I have no idea what it is. I write music as music without any reference other than to express

---

* See https://www.youtube.com/watch?v=hDRi-nGDw6AHD and https://www.youtube.com/watch?v=sTIVqzOK3MI

my feelings in tone.' As for Einstein, he once described Schoenberg's music as 'crazy'.

The twelve-tone method isn't really crazy, of course – just different. If a song is written in C major, then everything revolves around the seven notes of that scale. It can include notes from outside the scale, but only as 'accidentals' that gain their expressive effect from their relationship to C. Even within the scale, each individual pitch class has its own unique flavour. C, being the note that starts and ends the scale, is obviously the most important of all. The note a fifth above it, G, is called the 'dominant', which shows that it too is considered very important. And all the other PCs, such as the 'leading note' B, just a semitone below C, have specific roles to play. In contrast, the *relative* intervals between notes, such as that semitone between B and C, or the fifth between C and G, are secondary aspects that follow on from the *absolute* values of the notes themselves.

In contrast, Schoenberg's method focuses almost exclusively on the *relative* values of the twelve PCs – i.e. the intervals between them – rather than their absolute values. At this point you might object that, by definition, the interval between any two consecutive PCs is always a semitone, which isn't going to make very pleasant music at all. But Schoenberg's trick was to rearrange the twelve PCs into what he called a 'tone row', which then becomes the germ from which a musical work is constructed. To see how this works, let's look at the tone row that Stravinsky used for 'The Owl and the Pussy-Cat'.

As shown in the diagram, the row can take several different forms. The basic form is called the prime row, which by convention is taken as starting on PC 0, although the row (just like a major or minor scale) can be transposed up or

Prime row:	0	2	9	11	8	6	5	7	10	1	4	3
(in C)	C	D	A	B	Ab	Gb	F	G	Bb	Db	E	Eb
Inverse row:	0	10	3	1	4	6	7	5	2	11	8	9
(in C)	C	Bb	Eb	Db	E	Gb	G	F	D	B	Ab	A
intervals	2	5	2	3	2	1	2	3	3	3	1	3
Retrograde row:	0	1	10	7	4	2	3	5	8	6	11	9
(in C)	C	Db	Bb	G	E	D	Eb	F	Ab	Gb	B	A
Retrograde inverse:	0	11	2	5	8	10	9	7	4	6	1	3
(in C)	C	B	D	F	Ab	Bb	A	G	E	Gb	Db	Eb
intervals	1	3	3	3	2	1	2	3	2	5	2	3

**Details of the tone row used in Stravinsky's
setting of 'The Owl and the Pussy-Cat'.**

down by any number of semitones, giving twelve possibilities
in all (as an example, I've shown how the row comes out
when it starts on the note C).

The second form of the row, called the inverse, is just an
upside-down version of the prime row. At each step, instead
of going up so many semitones, you go down by the same
number of semitones. That means the intervals, or more
strictly the interval classes (which you'll remember only run
from one to six), between consecutive notes are the same in
both the prime and inverse rows. The sequence of intervals
is carefully chosen to give the sound the composer wants.
In this particular example, you'll notice that Stravinsky has
avoided the tritone (PC 6) and, more surprisingly, the major
third (PC 4).

Next comes the retrograde row, which, as the name suggests, is just the prime row backwards. Or rather, it starts
on PC 0 and reverses the sequence of intervals, so in this
case you'd have to transpose it up by three semitones to get
the literal reverse of the prime row. Finally comes the retrograde inverse, which you can either think of as the inverse

of the retrograde or (suitably transposed) the retrograde of the inverse.

All in all, there are 48 versions of the tone row: four forms, each in twelve possible transpositions. As you're free to jump around between all those different versions, this gives you a lot more freedom than you would have in a traditional key like C major. The thing that survives all that jumping around is the sequence of intervals, which can appear either horizontally, in a melody, or vertically in chords. Ultimately, it's this interval sequence that characterises the piece in the way that a key does in traditional Western music.

So, the twelve-tone system, devised by that arch-rulebreaker Schoenberg, does have rules of its own. But like all musical rules, they're not unbreakable. There's a great anecdote about a seminar Schoenberg gave to a group of students, explaining how his newly developed method applied to a piece of music he'd just written. One of the students, the composer Anton Webern, noticed that Schoenberg appeared to break his own rules at one point and asked why. 'Because it sounds better,' was Schoenberg's response.

Ultimately, that's the only way to judge any piece of music – by how it sounds. You might want to track down a good recording of 'The Owl and the Pussy-Cat' (such as the 'official' Stravinsky-approved one: https://www.youtube.com/watch?v=y4g1NYHBDSU) and make up your own mind. Personally, I think that, as songs about seafaring anthropomorphic animals go, it sounds miles better as a twelve-tone piece than it would have done in C major.

After Schoenberg, the musical avant-garde went off in two (not always mutually exclusive) directions. One was the exploration of electronic sounds, which we'll look at in detail in the next chapter. The other involved a highly mathematical extension of the twelve-tone method called 'serialism'. Just as the heart of a twelve-tone composition, the tone row, is an ordered set of PCs, so serial composition is based on ordered sets (or 'series', from which the name serial comes) of the whole gamut of musical parameters. These include everything from the pitch, duration and loudness of notes to changes of tempo and time signature and even which particular instruments are playing.

To confuse matters, some people apply the term 'serialism' retroactively to Schoenberg-style twelve-tone works, not just the later multi-parameter compositions that emerged in the 1950s and 1960s. Personally, I find this misleading, at least in terms of the music's effect on me as a listener. With the exception of a few composers like Webern, I don't feel that twelve-tone music per se sounds particularly 'modern' or 'experimental'. Many of Schoenberg's works from the 1920s onwards, for example, sound more restrained and 'neoclassical' than the lavish excesses of post-Wagnerian romantic music.

On the other hand, it's very difficult to do multi-parameter serialism – or 'total serialism', as it's sometimes called – in a way that doesn't sound out-and-out experimental. While there are plenty of twelve-tone works that I find genuinely moving and expressive, the best I can say about total serialism is that some of it is really *interesting* (perhaps the most damning adjective you can apply to an artistic work). The need to group musical parameters into ordered sets, and then manipulate those sets concurrently, turns the composition process into a complex intellectual problem, like a game of three-dimensional chess. That doesn't sound like a promising recipe for a great work of art.

That said, some composers, through sheer musical genius, have succeeded in making total serialism work in a way that transcends its mathematical complexities. By coincidence or otherwise, two of the most obvious names in this context are ones we'll meet again in the chapter on electronic music: the German Karlheinz Stockhausen and the American Milton Babbitt. The first of these was mentioned back at the start of this chapter as one of the people who introduced the term 'parameter' to music. Of his serial works, the one I like best – both because it sounds great and because it's conceptually easy to understand – is *Mantra*, an hour-long work for two pianos written in 1970. It's essentially the same thirteen-note phrase repeated over and over again but constantly expanded and contracted both horizontally and vertically so as to produce an ever-changing sound. As the title suggests, the overall effect is hypnotic and almost mystical.

As for Milton Babbitt – he was responsible for what has to be the most 'accessible' serial composition of all. Written in 1957 and scored for a jazz band, it lasts just eight minutes and doesn't sound like anyone's idea of 'mathematical music'. Yet its structure is firmly rooted in set theory – a fact that Babbitt jokingly referred to in the title of the piece: 'All Set'.

Algorithmic composition

At this point you may be feeling a little short-changed. The title of this chapter is 'Musical Algorithms', but so far, we've only talked about the broad-brush rules used by composers: things like time signatures, scales and the twelve-tone method. They aren't the sort of highly prescriptive rules people normally mean when they use the word algorithm. Yet, on occasion, musicians have employed those too. Let's take a brief dip into the rather scary world of species counterpoint.

As a basic concept, there's nothing particularly daunting about counterpoint. It just means that, rather than having accompanying instruments playing block chords, the harmonies are produced by interweaving multiple melodies together. Think of the opening of Led Zeppelin's 'Stairway to Heaven', where the acoustic guitar, playing one tune, is joined by English flutes playing a different tune. To modern ears this creates a distinctly 'olde worlde' effect because this kind of counterpoint was far more prevalent in the past than it is today. In the 17th and 18th centuries, in particular, it had a very strict set of rules – bordering on what might genuinely be called an algorithm.

In those days, students of composition were given exercises known as 'species counterpoint', where the teacher would specify a sequence of bass notes and the student would have to write a melody above it which obeyed a whole string of complicated rules. If you want to, you can try your hand at a simple 'first species' exercise online – but be warned that the computer (like an 18th-century music teacher) is very intolerant of having its rules broken.*

The thing about this kind of counterpoint is that, once you know rules, you don't need to use your brain at all. As the practice started to die out in the second half

---

* 'First Species Counterpoint', ToneSavvy, https://tonesavvy.com/music-practice-exercise/241/first-species-counterpoint-game/

of the 18th century, this was something musicians made quite a joke of. They developed a fashion for 'musical dice games', or *Musikalisches Würfelspiel* in German. As the name suggests, these used dice throws to put a musical piece together from a set of basic fragments. One such game, with the amusing title of 'Method for Making Six Bars of Double Counterpoint at the Octave Without Knowing the Rules', was produced by the composer C.P.E. Bach – who just happened to be the son of one of the all-time masters of counterpoint, J.S. Bach.

Even Mozart got in on the act, or so it's alleged. A couple of years after his death, his publisher put out one of these musical dice games with Mozart's name on the title page – and who are we to suggest he was just cashing in on the great man's reputation?

Premiere Partie. Erster Theil.

	A	B	C	D	E	F	G	H
2	96	22	141	41	105	122	11	30
3	32	6	128	63	146	46	134	81
4	69	95	158	13	153	55	110	24
5	40	17	113	85	161	2	159	100
6	148	74	163	45	80	97	36	107
7	104	157	27	167	154	68	118	91
8	152	60	171	53	99	133	21	127
9	119	84	114	50	140	86	169	94
10	98	142	42	156	75	129	62	123
11	3	87	165	61	135	47	147	33
12	54	130	10	103	28	37	106	5

Seconde Partie. Zweiter Theil.

	A	B	C	D	E	F	G	H
2	70	121	26	9	112	49	109	14
3	117	39	126	56	174	18	116	83
4	66	139	15	132	73	58	145	79
5	90	176	7	34	67	160	52	170
6	25	143	64	125	76	136	1	93
7	138	71	150	29	101	162	23	151
8	16	155	57	175	43	168	89	172
9	120	88	48	166	51	115	72	111
10	65	77	19	82	137	38	149	8
11	102	4	31	164	144	59	173	78
12	35	20	108	92	12	124	44	131

Two tables from the dice game attributed to Mozart; the idea is that the player chooses which order to put a set of musical fragments in by throwing dice and reading the appropriate 'IDs' from the tables.

https://gbrachetta.github.io/Musical-Dice/

The idea of adding a random element to music, whether by throwing dice or some alternative process, isn't quite as frivolous today as it was in Mozart's time. With the advent of movie soundtracks in the 20th century, together with the increasing use of ambient background music, new genres have emerged where the specific form of a work is less important than its general texture. From the 1950s onwards, several composers have experimented with so-called 'aleatoric music', after the Latin word for dice, *alea*. Stockhausen, who dabbled in aleatoric music for a time, likened it to the way you could move all the leaves on a tree around and it would still look like a tree of the same species. Similarly, changing the arrangement of notes in a densely textured piece of music – within appropriate limits – doesn't change the basic mood of the piece.

Aleatoric music opens up new possibilities when it comes to compositional 'algorithms'. One particularly adventurous musician in this respect was the American John Cage. He was an adherent of Zen Buddhism, which encourages spontaneity in all things – and there's nothing more spontaneous, when it comes to composing music, than letting a pair of dice make all your decisions for you. Cage really did do that on occasion, although one of his best known aleatoric works, *Music of Changes*, uses a different method more in keeping with his Eastern philosophy. All the musical parameters – pitch, duration, loudness, tempo and so on – were chosen from tables of 64 options, the one to use being selected via an ancient Chinese divination method called the *I Ching*, which uses a series of coin tosses to produce a random number between one and 64.

Most fully algorithmic music involves some kind of randomness, whether from throwing a pair of dice, consulting the *I Ching* or – the most practical option – simply using a computer's built-in random number generator. But 'self-writing music', if that's the appropriate term, can be achieved in other ways besides invoking randomness. Composers can draw on other outside sources for the input to their musical algorithms. Another of Cage's works, *Atlas Eclipticalis*, takes its name from an astronomical atlas – which, in a sense, was a co-composer of the music. What Cage did was to lay semi-transparent sheets of music paper onto the pages of the atlas, then use the star positions to trace musical notes on the paper.

That's as extreme as algorithmic composition gets, leaving Cage with very little room to insert his own ideas or personality into the music. But maybe, as a Zen Buddhist, that's what he was aiming for. Other composers have made creative use of outside input in a way that leaves them more flexibility to do their own thing. Take Peter Maxwell Davies, for example. As Master of the Queen's Music from 2004 until 2014, he can hardly be described as a spaced-out hippie.* But he often drew, in a quasi-mathematical way, on pre-existing patterns in his music. One of his works, dating from 1975, shares its name with another, much older, work we encountered when we were talking about medieval modes: the plainchant 'Ave Maris Stella'. According to Davies, the medieval chant 'forms the backbone' of his own piece, 'projected through the magic square of the Moon'.

The 'magic square of the Moon' is a nine-by-nine array of numbers, arranged so that every row, column and diagonal adds up to 369. It's one of several such squares that have acquired astrological associations; in another piece from the

---

* Actually, I didn't say that about John Cage either, but I thought it.

same decade, *A Mirror of Whitening Light*, Davies used the eight-by-eight square of Mercury. In the case of 'Ave Maris Stella', he used two versions of the lunar magic square, one to provide pitches for his composition, the other to provide the rhythmic parameters. Having seen both squares, I must admit I can't see how they relate back to the original chant.* But I suspect this works in Davies' favour. Rather than rigidly applying a predetermined algorithm à la John Cage, it's more likely that he used his sources of inspiration – the plainchant and the magic square – as a trigger to stimulate his own creativity.

Whatever algorithm Davies used, it's almost certain that – back in the days when very few composers had access to a computer – he had to implement it manually, with paper and pencil. These days, algorithmic composition, along with every other aspect of music that has been transformed by technology, is a lot simpler than it used to be – as we'll see in the next chapter.

---

* In *Peter Maxwell Davies* by Paul Griffiths (London: Robson Books, 1985), p. 73.

THE ELECTRONIC AGE 4

The idea of creating 'technology-based' musical instruments has been around a long time; church organs, for example, had become fairly sophisticated pieces of engineering by the end of the 15th century. By the first decade of the 19th century, the German inventor Johann Maelzel was putting the finishing touches on an even more amazing gadget. This wasn't the invention for which he's best remembered – the metronome – but a huge machine he called the 'panharmonicon'. Powered by compressed air, it was a self-playing musical instrument (now sadly lost in history) that could simulate a variety of sounds from drums, trumpets and flutes to cannon and musket fire.

To promote his new invention, Maelzel commissioned the most famous composer of the day, Beethoven, to write a piece specifically for the panharmonicon. The result was Beethoven's fifteen-minute 'Battle Symphony', more formally known as 'Wellington's Victory'. Making full use of the available sound effects and incorporating popular tunes such as 'Rule Britannia!' and 'For He's a Jolly Good Fellow',

**An illustration of Maelzel's panharmonicon, a
kind of self-contained mechanical orchestra.**

Unknown author, public domain, via Wikimedia Commons

it's great fun – and, fortunately for panharmonicon-less pos-
terity, can be played by a conventional human orchestra too.
When a critic pointed out that it was, perhaps, a tad less
sophisticated than Beethoven's other symphonies, the great
composer came back with the immortal reply: 'What I shit
is better than anything you could write.'

As for 'electric instruments', the first viable one dates

from 1759. I'll pause a second to let that sink in: not 1959, or even 1859, but *1759* – the year George Frederick Handel died. It was called the *clavecin électrique*, and it used a standard musical keyboard to operate a series of tuned bells by altering their static electric charge. The first 'sound synthesiser' followed around a century later – not as a musical instrument, but a scientific one. Designed by the physicist Hermann Helmholtz, it used electromagnets to vibrate an array of tuning forks of different sizes. Because the sound produced by a tuning fork is close to a pure sine wave, Helmholtz was able to use his device to recreate more complex sounds made up of a number of partials. This reverses the process of Fourier analysis described in Chapter 2, and – as we'll see shortly – is an example of what in today's jargon is called 'additive synthesis'.

Actually, I was playing with words a little in the previous paragraph. There's a difference between the adjective 'electric', which simply means powered by electricity, and 'electronic', which relates to electrons – subatomic particles that weren't even discovered until the final years of the 19th century. While the *clavecin électrique* and the Helmholtz synthesiser are electric, they're not electronic – and it's only with electronics that really sophisticated technology becomes possible. It's worth taking a step back at this point to see just why that's the case.

Amplifiers and oscillators

Despite the similarity in names, electricity doesn't always involve the flow of electrons. There are tiny electric currents flowing in the nerves of your body, but they're carried by charged sodium and chlorine ions rather than free-moving electrons. It's true, however, that almost all electrical gadgets involve the flow of electrons through metal wires. That's not enough to make them 'electronic', though. By convention, that word is reserved for certain devices that explicitly control the electron flow.

It may seem like a roundabout route, but it will simplify the later discussion if we start with a few basic devices that are 'electric' but not 'electronic'. First, there's the electric motor, which passes a current through an electromagnet to turn a shaft. This process works both ways, so that turning the shaft by hand causes an electric current to flow, in the simplest kind of electrical generator. In a slightly less obvious way, the same two-way movement-to-electricity transformation explains how an old-fashioned, pre-electronic telephone works. An electromagnetic transducer in the mouthpiece converts sound vibrations into a rapidly varying electric current, while a similar transducer in the earpiece transforms electrical signals back into sound waves.

In parallel with these practical inventions, which all date from the 19th century, more academically minded scientists were experimenting with a completely different aspect of electricity. This involved passing a current through glass tubes containing either a vacuum or tenuous gas, which resulted in various phenomena that – apart from their abstract scientific interest – appeared to have no practical use at all. Yet, unbeknownst to anyone at the time, these experiments were laying the groundwork for the electronic revolution that would shape the future.

The first practical discovery, and the world's first genuinely 'electronic' device, was the diode valve. This contained two electrical connections, called the cathode and the anode, inside a glass vacuum tube, with the cathode being heated by a separately powered filament. When the valve was wired up so that the cold anode was at a positive voltage relative to the heated cathode, the latter emitted a stream of electrons (which are negatively charged particles). As a result, a steady electric current flowed through the diode.

At this point you may be thinking: 'So what?' After all, you could achieve the same effect, and a lot more simply, with a piece of copper wire. But think what happens if you connect the valve the other way around. Electrons can't flow from the anode because it isn't heated, and they can't flow from the cathode because it's now at a positive voltage. So, just like a one-way water valve, the diode allows electron flow in one direction while preventing it in the opposite direction. A diode can be used, for example, to convert the alternating current (AC) produced by most electrical generators into the direct current (DC) that's often more useful in practice.

The first commercial diodes hit the market in 1915, by which time an even more important device had been invented. You may not be surprised to learn that it was called a 'triode', and that it had a third electrical connection in addition to the diode's cathode and anode. But the way that third connection was used is a stroke of genius. Taking the form of a grid located between the two other electrodes, its voltage could be raised or lowered by an external source. The more negative the grid voltage, relative to the cathode, the less current flows through the tube – in other words, the current goes up and down in sync with the grid voltage. But the latter can be quite small – for example, the output from a microphone – while the current through the tube can be as huge as the power supply permits.

By playing around with three metal plates inside a vacuum tube, we've just built the world's first electronic amplifier. And more than that, if we think things through a little further. The amplifier only works if we're careful to keep the grid voltage within certain limits. If it becomes too negative, then all the electrons will be blown back to the cathode and no current will flow at all. So we've also made an electrically controlled switch – a way to turn an electric current on or off by applying one voltage or another to the control grid. In the 1940s, a few decades after the triode was invented, that's how people built the world's first digital computers.

The triode has a reasonable claim to being the single most ubiquitous invention of modern times. It's true that, unless you're a retro-electronics geek like me, you're very unlikely to own a single vacuum tube. But in terms of the basic functionality of a triode, I can say with reasonable certainty that you have several billion of the things in your possession. In the middle of the 20th century, it was discovered that a miniature, solid-state version of a triode could be made from crystalline silicon – and, over the course of time, it became possible to etch more and more of them onto a single chip of silicon. Under the name 'transistor' rather than triode, these have infiltrated modern life so thoroughly that we scarcely notice them. An iPhone 14, for example, contains around 16 billion transistors, while the transistor count for some specialist 'artificial intelligence' chips can exceed a trillion.

Returning to the subject of amplifiers: a good quality one, capable of producing a high-fidelity response over a wide range of frequencies and input levels, requires a lot more than a single triode – or a single transistor – but the basic principle is the same. It takes an input signal and transforms it into an output signal that is louder by a factor called the 'gain'. Amplifier technology has been indispensable to music over the past century, not only in faithfully reproducing broadcast or recorded music in our homes, but – in at least one important case – shaping the actual sound of the music itself. I'm talking about the electric guitar.

It's an 'electric' guitar, not an electronic one, for the simple reason that the instrument itself doesn't contain any electronics. It just has a set of electrical pickups, which work much like microphones to convert the acoustic sound of the guitar strings into an electrical signal that can be sent to an amplifier. But from then on, it's electronics all the way. If the amp is used the way it was originally intended – and the way amplifiers are used in most other contexts – then its role would be relatively minor, simply boosting the volume of the guitar without changing the basic quality of the sound.

That's what happens when you use a guitar amp on a clean, low-gain setting. But if you crank the gain up to maximum, then you're trying to amplify the guitar signal to a higher voltage than the power supply can provide you with. This results in 'clipping' of the signal, with the smoothly rounded tops of the waveform brutally chopped off – creating a harsher sound more like a square wave. In the frequency domain, this corresponds to much more prominent high-frequency partials than were present in the original signal – or, in a word, 'distortion'. To an electronics engineer, that's a bad thing, but to hard rock fans it's one of the best things that's ever happened to music. It transforms the electric guitar from what might have been a rather bland instrument into – in the hands of someone like Jimi Hendrix – one of the greatest virtuoso sounds of all time.

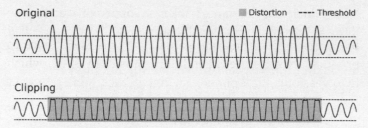

Simplified illustration of guitar distortion. A sine wave is reproduced perfectly if it remains below the amplifier's distortion threshold but is 'clipped' to something more like a square wave if it goes above it.
Mikhail Ryazanov, CC-BY-3.0

So, more often than not, the sound you hear from an 'electric' guitar is an electronic one rather than simply an amplified acoustic sound. That's even more the case if the guitarist uses one or more effects pedals, which really do have electronics inside them.

Another way to manipulate a guitar sound – and another Hendrix favourite – is through the use of feedback. Just like clipping-induced distortion, this takes what to lesser mortals might seem to be an undesirable artefact and turns it into a positive musical feature in itself. In a nutshell, feedback involves placing a guitar pickup – or any microphone-like device – close enough to the loudspeaker to pick up the already amplified sound and amplify it still further – in what would be an endless loop of ever-growing sound if there wasn't such a thing as a clipping threshold.

In fact, feedback doesn't even require an acoustic input to get going. You'll know that all too well if you've ever picked up the microphone of a PA system, only to be greeted by an ear-piercing squeal before you've even opened your mouth. The sound here is determined by what's known as the 'resonant frequency' of the system. It isn't something many people would describe as a pleasant or 'musical' sound, but it did inspire the avant-garde composer Steve Reich to produce a work called 'Pendulum Music' in 1968, which involves four microphones swinging back and forth over four loudspeakers.

As raucous as it sounds when it occurs accidentally, it's possible to tame this kind of 'no-input' feedback and put it to good use. If you feed the output of an amplifier – I mean an amplifier chip on a circuit board, not a guitar amp – back into its input via a suitably designed filter, then you can create a stable sound with a waveform that oscillates steadily at a frequency determined by the filter circuit.* The result, not too surprisingly, is called an 'oscillator' – and it's the basis for almost all the electronic music ever produced.

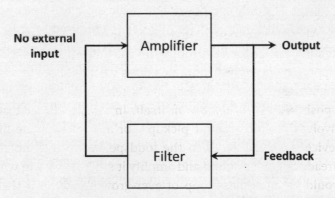

An amplifier can be turned into an oscillator by feeding its output back into the input via a suitably designed filter.

---

* In audio processing, a 'filter' is a circuit that removes unwanted features from a signal, analogous to the way a car's air filter removes dust particles from the air inflow.

One of the first practical uses of oscillators was in radio circuits, where they operate at frequencies much higher than the audible range. In the 1920s, a French engineer named Maurice Martenot noticed that if two radio oscillators are tuned to slightly different frequencies, they produce an audible tone at the difference between those two frequencies – by a process analogous to the 'beating' of two slightly out-of-tune notes mentioned in Chapter 2. As Martenot was something of an amateur musician, he felt this rather spooky effect – known in radio circles as 'heterodyning' – was too good to go to waste. He ended up designing a whole new musical instrument around it, which came to be known as the 'ondes Martenot', from the French for 'Martenot waves'.

It's fair to say that, despite its unique sound, the ondes Martenot never established much of a foothold in the world of music. The only composer who really took it to heart was another Frenchman, Olivier Messiaen, who used it, among other places, in his greatest work, the *Turangalîla-Symphonie* of 1949. Even if you don't feel up to the entire 76 minutes of it, I'd recommend listening to the first couple of minutes just for the symphony's weird mixture of electronic and orchestral sounds. It's a startling effect, which lifelong Messiaen fan Jonny Greenwood tried to emulate on several Radiohead tracks, most notably on the *Kid A* album. Greenwood even used an ondes Martenot when recording the song 'How to Disappear Completely'.

If you do sample the *Turangalîla-Symphonie*, you'll notice that a distinctive feature of the ondes Martenot is its use of what musicians call 'glissando' – sliding continuously between notes rather than jumping from one to the other. Although the instrument has what looks like a piano keyboard, this doesn't actually do anything – it's just there to give the player an indication of where the different pitches are. The notes are actually produced by sliding a metal ring along a horizontal wire running in front of the keyboard.

If that sounds weird, it's actually pretty conventional compared to the playing technique of the theremin, another early electronic instrument that uses the same 'heterodyning' method of producing notes. Invented by Russian physicist Leon Theremin, this instrument doesn't require any physical contact to be played. If you're old enough to remember when portable radios had long telescopic antennas, you may also remember that the sound sometimes altered if you moved your hand close to the antenna without actually touching it. That's basically how a theremin is played; it has two antennas, one controlling pitch and the other volume, that are able to sense the position of the player's hands through their effect on an electrical property called capacitance.

The sound of the theremin is as weirdly futuristic as its playing technique – to the extent that it became a de rigueur feature of sci-fi movie soundtracks in the 1950s. You can hear it, for example, in *The Day the Earth Stood Still* (1951), *The Thing from Another World* (1951) and *It Came from Outer Space* (1953). Amusingly, however, the best known 'theremin' sound of all – the theme tune of the original 1960s *Star Trek* series – isn't a theremin at all, but a female singer imitating one.

In hindsight, although the theremin and ondes Martenot sound irrefutably 'electronic', they don't really equate to what we think of as electronic music today. That's because you can't characterise modern electronic music with any one single *sound*. Its genius lies in its inexhaustible variety – not just of weird and wonderful waveforms, but the way whole libraries of sounds, whether they are electronically produced or sampled from acoustic music or nature, can be edited together to create a mind-blowing tapestry of sound.

One of the great pioneers of this aspect of electronic music was Edgard Varèse, whose 'Poème Électronique' was mentioned earlier. Created on magnetic tape, this was played through more than 400 loudspeakers inside the Philips Pavilion at the 1958 World Fair in Brussels. It consists of a montage of short, pre-recorded sound clips and pure electronic sounds, produced by Varèse in collaboration with the research division of the Philips electronics company.

With a captive audience at the World Fair, 'Poème Électronique' would have reached far more listeners than most avant-garde music of its time. Even so, it pales beside another pure-electronic work of the same decade: the jarringly innovative soundtrack of the 1956 sci-fi movie, *Forbidden Planet*. Starring Leslie Nielsen as a kind of proto-Captain Kirk, and with a plot loosely based on Shakespeare's *Tempest*, the film would have been memorable enough anyway, but it's the avant-garde score – like nothing Hollywood had ever heard before – that ensures its immortality.

Consisting of nothing but electronically produced sounds, the score was created by the husband-and-wife team of Bebe and Louis Barron. They'd cut their musical teeth as assistants on one of John Cage's aleatoric compositions – *Williams Mix*, named after the African American architect Paul Williams who commissioned it. The Barrons' role in that instance was to create a library of short sound clips, both recorded natural sounds and electronically generated ones, which were passed to the composer to be spliced together. Cage then threw dice – literally – to decide which order to put the clips in.

The *Forbidden Planet* score, on the other hand, was the

work of the Barrons alone. Unlikely as it may sound, they took inspiration from a textbook on cybernetics by Norbert Wiener. This was something of a bible for electronics engineers, and the Barrons adapted many of the feedback circuits described by Wiener to produce a whole spectrum of unearthly sounds. These were recorded onto magnetic tape, after which they were manipulated in various ways – by splicing, looping and changing the tape direction and speed – and repeatedly rerecorded. In the end, it's difficult to say if the result is music or just a compilation of weird sound effects. To perpetuate the ambiguity, *Forbidden Planet*'s score is described in the closing credits not as music, but as 'electronic tonalities' – albeit for legal rather than aesthetic reasons. The producers wanted to avoid trouble with the Hollywood musician's union, which the Barrons didn't belong to.

It was to produce electronic sound effects, rather than music, that the BBC's pioneering Radiophonic Workshop was set up in 1958. Despite this, its best-known creation was a piece of music – and once again it was associated with the sci-fi genre. I'm referring, of course, to the theme tune for the *Doctor Who* TV series, which was laboriously pieced together in 1963 by Radiophonic Workshop composer Delia Derbyshire. She employed a mixture of sine-wave and square-wave oscillators, a noise generator and a beat-frequency generator. As natural and flowing as the result sounds, it was pieced together virtually note by note, followed by weeks of cutting, shaping and filtering before the individual tape tracks were ready for synchronisation and mixing.

This is a far cry from the situation today, where, with the aid of a DAW and a suitable set of plugins, the same processes could be got through in an afternoon. In the early days, however, producing electronic music was a slow and difficult process, and only a few pioneering specialists like Varèse, the Barrons and Derbyshire ventured anywhere near it. The obvious question at this point is 'why?' – what was it about electronic music that attracted people to explore its possibilities, when doing so was such painfully hard work? The answer lies in the virtually unlimited range of previously unheard sound textures – or 'timbres', in musical jargon – that can be created electronically. Thanks to technology, and the science that underlies it, music was no longer limited to the kind of sounds that could be produced by an acoustic instrument.

From science to art

The key to producing pure electronic timbres is a basic scientific principle we met in Chapter 2, namely Fourier analysis. We saw there how an acoustically generated sound wave can be broken down into a series of single-frequency sine waves. By reversing the procedure, it's possible to add a number of sine tones together to produce a more complex sound. Linguistically, the opposite of analysis is synthesis, so what we've done here is to create a 'synthesiser' – the most flexible of all electronic instruments. Actually, we've already met one of these in the form of the 19th-century Helmholtz synthesiser. That used tuning forks rather than electronically produced sine waves, but the principle is exactly the same. And because it involves adding simple waves together to create more complex ones, it's called 'additive synthesis'.

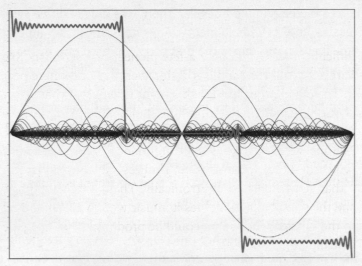

**In additive synthesis, a number of sine waves
(thin black lines) are combined to produce
a more complex waveform – in this case, an
approximation to a square wave (thick grey line).**

This diagram shows how, by combining a large
number of sine waves, additive synthesis can produce
something closer to a square wave. This has a much richer
musical timbre due to the large number of high-frequency
partials – more like a clarinet than a tuning fork. In practice, however, this kind of additive synthesis isn't the best
way to produce complex sounds because so many individual
sine waves are needed to produce the required partials.
It's much simpler to build an oscillator that produces a
square wave to start with, and then use filters to subtract
any unwanted partials. That's the approach used in most
modern synthesisers, and – surprise, surprise – it's called
'subtractive synthesis'.

In the beginning, however, there was only additive synthesis – and, as with so many things in 20th-century music, Stockhausen was one of the first people to try it out. In 1953, he produced *Studie I*, which used individually tailored 'notes' synthesised from up to six basic sine tones. It was the first – but far from last – time in musical history that a composer had complete control over timbre, rather than being forced to take whatever an instrument could provide.

The following year, in *Studie II*, Stockhausen went even further. He realised that, with the whole frequency range at his disposal, he no longer had to use the twelve notes of a standard octave scale. So, he created his own scale, based not on the octave – a frequency ratio of two – but on a frequency ratio of five. Just as an octave is divided into twelve equally spaced semitones, Stockhausen divided his new scale into 25 equal parts – which ended up about 10 per cent larger than a semitone. That may seem a crazy thing to have done, but it was really just a way of expressing the amazing sense of limitless freedom that electronic music offered.

With synthesisers, and their counterparts in computer software, being so widespread today, it's difficult to appreciate just how trailblazing Stockhausen's work was in the 1950s. There was simply no precedent for what he was doing in the musical world – only in the scientific one. As he recalled later:

> I started looking in acoustic laboratories for sources of the simplest forms of sound wave, for example sine wave generators, which are used for measurement. And I started very primitively to synthesise individual sounds by superimposing sine waves in harmonic spectra, in order to make sounds like vowels: *aaah, oooh, eeeh* etc., then gradually I

found how to use white-noise generators and electric filters to produce coloured noise, like consonants: *ssss, sssh, fffh* etc. And when I pulsed them it sounded like water dripping.

Adapting scientific equipment to create new musical sounds was just the first step in what Stockhausen wanted to do. He then had to put all those sounds together to create a work of music, such as his 35-minute composition *Kontakte*. Despite dating from 1960, this still sounds fresh and 'modern' if you listen to it today. But unlike contemporary electronic music, putting *Kontakte* together was a painfully slow process – as Stockhausen explained:

> In some sections of *Kontakte* I had to splice it all by hand, which is an unbelievable labour. Imagine, I worked on the last section of *Kontakte* together with Gottfried Koenig in studio eleven on the third floor of Cologne Radio for three months. And when it was completely ready, I spliced it together with the previous sections, listened, turned pale, left the studio and was totally depressed for a whole day. And I came back next morning and announced to Koenig that we had to do it all over again. I mean, he almost fainted.*

Stockhausen, like Delia Derbyshire at the BBC Radiophonic Workshop, created music in a laboratory-like studio that was filled with oscillators, amplifiers and other pieces of electronic equipment. But around the same time, on the other

---

* The two Stockhausen quotes are from Robin Maconie, *Stockhausen on Music* (London: Boyars, 1989).

side of the Atlantic, another approach was being developed. Why not put all that diverse functionality into a single, all-singing (but not necessarily all-dancing) contraption?

The first electronic device to bear the name 'synthesiser' was built by the RCA company in 1955. Located at the Columbia-Princeton Electronic Music Center in New York, it was a monstrous thing that effectively packed an entire Stockhausen-style studio – oscillators, filters, noise generators and all – into a single package. And it didn't stop there. While Stockhausen had to operate everything by hand, the RCA synthesiser could be programmed – or 'sequenced', in modern jargon – using a roll of punched paper tape, in the style of the giant, vacuum-tube-packed computers of the same era.

A one-off, the RCA synthesiser was basically a research tool that was much too complicated and temperamental for anyone to make hit records with. As far as musicians are concerned, the only one who had sufficient patience to use it was another of the experimental composers we met in the previous chapter, Milton Babbitt – of 'All Set' fame. His best-known synthesiser piece, 'Philomel', dates from 1964 and is based on a Greek legend about a princess who is transformed into a nightingale. It's sung by a soprano, who is accompanied both by the synthesiser and by an electronically processed tape of her own voice. Here's how Babbitt described the RCA machine many years later:

That word 'synthesizer' connotes some little boy with a small box, sitting at a keyboard. That was far from the case. This was a programmed instrument that was more than the length and size of this room, so people saw it and they thought it was a computer, although it wasn't.

> It couldn't compute anything, did no number-crunching,
> and had no memory – for which it was probably grateful.
> This particular 'computer' was mine, more or less, because
> nobody else but I had worked on it. I went down there and
> worked, and I knew how to use it. It was a very recalcitrant
> instrument; you had to do everything yourself, it was very
> hard. You programmed every aspect of a musical event
> and the mode of progression to the next event. Then you
> recorded it on tape.*

As Babbitt says, there is no sense in which the RCA synthesiser was a computer. It's true that it was 'programmable', but
so was Maelzel's panharmonicon, or an old-style piano-roll
player. The essence of a computer is that it turns whatever
data it's given into numbers – or 'digits', hence the word digital – and then manipulates those numbers without needing
to know what they represent. It might be a sound wave, or
the image of a cat, or a shopping list – it's all the same to
the computer. On the other hand, analogue devices like the
RCA synthesiser represent sounds in something quite close
to their acoustic waveforms – simply using electrical voltages
in place of air pressure levels.

It may come as a surprise, but the origins of digital music
can be traced back to the 1950s – the same decade that saw
Stockhausen and Babbitt locked in heroic battles with their
primitive analogue equipment. One of the key figures in
the development of computer music was a multi-talented
engineer named John R. Pierce, who was employed at Bell
Laboratories in New Jersey. You may never have heard his
name, but that can hardly be said of one of the technical

---

* From the liner notes to the 'All Set' CD (BMOP/sound, 2013).

terms he coined: transistor. He came up with that word by analogy with two earlier solid-state devices: the thermistor and the varistor.

Under the amusing pen name of J.J. Coupling, Pierce wrote a series of non-fiction articles that appeared in the magazine *Astounding Science Fiction* in the 1940s and 1950s.* One of these, from November 1950, provides an early example of his interest in the science of music. Called 'Science for Art's Sake', it describes the concept of aleatoric composition – which Pierce approached in the same spirit as C.P.E. Bach and Mozart: 'By the throwing of three especially made dice and by the use of a table of random numbers, one chord was chosen to follow another.'

Later in the 1950s, Pierce and his Bell Labs colleague, Max Mathews, developed the first computer programs for sound creation. Sadly, although the music they produced was technically very impressive, it never gained much attention outside the scientific community. This despite the fact, as Mathews wrote in 1963, 'There are no soldering irons, tape splicings or even knob-twistings involved, as there are with other electronic equipment for producing music.' Pierce even made a 10-inch demo record, *Music from Mathematics*, which he sent to prominent musicians of the day such as Leonard Bernstein and Aaron Copland – with little result other than polite brush-offs.

For all that, one of Mathews and Pierce's digital compositions did attain immortality of a kind. In 1961, they used a digital speech synthesiser to produce what must be the first 'electronic vocal' work in history – a rendering of

---

* OK, it's probably only amusing to people who studied physics at university. It's just that 'j-j coupling' is one of the types of spin-orbit interaction that occurs inside atoms.

the Victorian popular song 'A Bicycle Built for Two'. When Pierce's friend Arthur C. Clarke heard it a few years later, he was so impressed that he persuaded Stanley Kubrick to include it in the movie version of *2001: A Space Odyssey*. It's what the psychotic computer HAL 9000 sings when he's being powered down.

As to the mechanics of computer music, as good a description as any can be found in a textbook, *The Science of Musical Sound*, that Pierce wrote in 1983:

> How is it possible for a computer to generate sounds? The sampling theorem gives us a clue. Consider any waveform made up of frequency components whose frequencies are less than B. That is, consider any sound wave whose frequency components lie in the bandwidth between zero and B. Any such waveform or sound wave can be represented exactly by the amplitudes of 2B samples per second. These samples are merely the amplitudes of the waveform at sampling times spaced 1/2B apart in time.

The only problem here is that the word 'sampling' may conjure up a misleading image in many people's minds. Maybe you think of a two-second snippet of, say, a drum pattern. If it was recorded with an old-fashioned analogue tape recorder, that's more or less what it is – a two-second snippet of sound (or an electronic analogue of it). But in the digital case, it's not really that at all – it's just a string of numbers. As Pierce says, these numbers simply measure the amplitude of the acoustic signal at regularly spaced intervals. And to an engineer, it's those regularly spaced measurements that constitute the 'samples', not the clip as a whole.

What the sampling theorem says is that, to faithfully record a sound that has a bandwidth of B Hz, the sample spacing has to be 1/2B seconds. The standard bandwidth of CD-quality recording is 22,050 Hz, meaning that we need 44,100 samples every second if we're going to hear all the frequencies we need (except oldsters like me, who struggle to hear anything above 10 kHz).

So far in this chapter, all the electronic music we've encountered has come from the 1950s and early 1960s. We've been talking about musicians and experimenters – Varèse, Stockhausen, Babbitt, Bebe and Louis Barron, Delia Derbyshire, Max Mathews, John Pierce – whose names are virtually unknown to the public at large. Yet between them, they developed almost all the basic techniques that are used by a present-day musician sitting in front of a computer screen. The only difference is in the way those techniques were accomplished. Here's a quick rundown of the sort of techniques I'm talking about:

- Mixing of several recorded tracks into one, by playing them back simultaneously and re-recording the combined sound.
- Modifying the signal in the frequency domain, using electronic circuits called 'high-pass' and 'low-pass' filters, which as the name suggests only pass frequencies above or below a certain threshold. An equaliser, as mentioned briefly in the first chapter, uses a bank of filters to tailor a sound across the whole frequency spectrum.
- Shaping the way the volume of a waveform changes as it's played, technically referred to as its attack-decay-sustain-release (ADSR) envelope. It's easiest to understand this in terms of what happens when you hit a piano

key ('attack' followed by 'decay'), then hold it down ('sustain') and finally let go ('release').

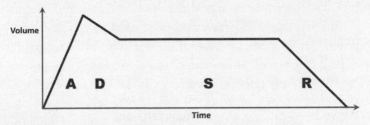

The overall shape of a synthesised sound is characterised by its ADSR envelope.

- Adding reverberation, or 'reverb' – which is something that happens naturally when acoustic music is played in a resonant space like a concert hall but has to be added artificially in electronic music.
- Altering the speed and/or direction of the waveform – a relatively straightforward operation even in the days of tape recorders, as long as they had a 'varispeed' capability.
- 'Looping' a sound clip, which is such a familiar concept in contemporary music that it's difficult to remember that at one time it literally involved a closed loop of tape going round and round over the playback head of a tape recorder.
- Adding echo or delay effects, which originally called for two separate tape heads, one to record and the other to play back over a loudspeaker. As the tape went through the recording head, it overdubbed the sound from the speaker (playing an earlier part of the tape) onto a later section.

Several of these effects were popularised by The Beatles – for example in 'Strawberry Fields Forever', which was discussed in Chapter 2. But it's a mistake to imagine that The Beatles invented them – or even that they unknowingly 'reinvented' techniques that had previously been used by classical composers. Philip Norman makes this clear in his 2016 biography of Paul McCartney. He explains how Barry Miles, the proprietor of a trendy London bookshop in the 1960s, introduced Paul to the music of Stockhausen, Varèse and – a name we haven't encountered yet – the Italian composer Luciano Berio. This was an eye-opener to Paul, and something his fellow Beatle John Lennon had difficulty keeping up with. As Norman explains:

> Whenever Paul saw John, he'd be full of news from this esoteric new world: how the tape-loop maestro Luciano Berio was coming to Britain to teach a course at Dartington Hall; how Paul had written a fan letter to Stockhausen – quite a turnaround for a Beatle – and received a personal reply. 'God, man, I'm *so* jealous,' John would respond gloomily, but always left it at that.

Stockhausen, in particular, had a strong impact on the world of popular music – all the more surprising given that even his most ardent devotees have to admit his work is 'difficult'. You may never have heard a note of Stockhausen's music, but there's a fair chance that one or other of your favourite artists were influenced by him at some point. If you extend that to the next degree of separation – musicians who were influenced by musicians who were influenced by Stockhausen – then it becomes pretty much a certainty. Besides Paul McCartney, Stockhausen has been name-checked by (deep breath) Frank Zappa, Miles Davis, Grateful Dead, The Doors, The Who, Pink Floyd, David Bowie, Brian Eno, Tangerine Dream, Kraftwerk, Orchestral Manoeuvres in the Dark, Sonic Youth, Björk, Portishead, Radiohead, Venetian Snares – and probably many others I'm not aware of.

So, the question is – how did this weird-sounding, highbrow activity of the mid-20th century turn into the worldwide phenomenon we know and love today? Let's pick up the story where we left off in the mid-1960s.

The electronic revolution

The turning point came in 1964, when the American engineer Robert Moog developed a modular analogue synthesiser that was compact and practical enough that it could be built in quantity and marketed commercially. That's to say, it was 'compact' and 'practical' when compared to Milton Babbitt's room-filling RCA synthesiser. The first Moog synths were still pretty huge by modern standards, consisting of separate modules containing oscillators, filters and amplifiers that the user had to wire together using telephone-exchange style patch cables. Even so, it was a huge step forward for electronic music, taking it out of the realm of experimental avant-garde music and into the every day world of rock and pop.

One of the first rock bands to employ a Moog synthesiser was The Doors on their second album *Strange Days* from 1967. You can hear it clearly both in the title track and in the distinctly avant-garde 'Horse Latitudes'. The band were introduced to the synth during the recording sessions by one of its early adopters, Paul Beaver. Here's how Doors keyboardist Ray Manzarek described the moment in his auto-biography, *Light My Fire*:

> Paul Beaver brought his huge modular Moog system into the studio and began plugging a bewildering array of patch cords into the equally bewildering panels of each module. He'd hit the keyboard and outer space, bizarre, Karlheinz Stockhausen-like sounds would emerge. He would then turn a mystifying array of knobs placed in rows around the patch cord receptacles and more and different space would emerge.

The Moog is just one of the instruments used on *Strange Days*, alongside drums, guitar and Manzarek's own combo organ, keyboard bass and marimba. It wasn't long, however, before the synth was starring on its own. The breakthrough came with the 1968 album *Switched-On Bach* by Wendy Carlos (known at the time as Walter Carlos). Carlos was a musical and electronic wizard who, in 1956, at the age of just seventeen, had built an electronic music studio and then went on to form a collaboration with Robert Moog ten years later. The aim of *Switched-On Bach* was to demonstrate the Moog's potential via electronic arrangements of works by J.S. Bach. That could have been a risky undertaking, given that it was close on 250 years since Bach had represented the peak of musical fashion – yet it paid off. *Switched-On Bach*, featuring tracks such as 'Air on a G-string', reached number ten in the US album charts, and the day of the synthesiser had well and truly arrived.

Robert Moog's next step was to create an instrument suitable for live performances on stage – a single self-contained unit with user-friendly knobs and switches instead of clunky patch cords. The result was the Minimoog – a masterpiece of compact design that first hit the market in 1970.

A Minimoog synthesiser from the 1970s.
Wikimedia user Qwave, CC-BY-SA-4.0

The Minimoog was so elegantly laid out that we can use a photograph of one as a kind of 'flow chart' to understand how an analogue synth works. One part that even Bach would have recognised is the musical keyboard, which is played just like an organ or piano. To the left of this are two wheels, labelled 'pitch' and 'mod'. The pitch wheel raises or lowers the pitch being played, rather as a guitarist can do by bending a guitar string. The mod – short for modulation – wheel also affects the pitch being played, but it does so on a regularly repeating pattern, a bit like an opera singer using vibrato.

Above the keyboard is a series of panels which are best read from left to right. First, the controllers set basic things like the tuning of the instrument, the amount of glide from one note to the next and the type of modulation used by the mod wheel. Next come the oscillators, three of them, which produce the synth's basic sounds, in the form of sine waves, sawtooth waves or square waves. These sounds are then combined, at the desired levels, in the mixer section, where it's also possible to add a certain amount of 'noise' (which, in a technical context, means the kind of static you hear from an untuned radio). Finally, the 'modifiers' section contains a filter that can remove high-frequency partials, and knobs to shape the attack-decay-sustain-release contour.

The Minimoog proved enormously popular, with over 10,000 of them sold through the 1970s. Today, they're expensive and highly sought-after museum pieces, but if you happen to have an iPad you can buy the next best thing – Moog Music Inc.'s 'Minimoog Model D Synthesizer' app.* It even boasts a few features, such as a range of pre-loaded settings, which the original didn't have.

Inspired by Moog's success, other companies started to go into the synthesiser business. Early entrants into the market were Alan R. Pearlman's ARP Instruments in the United States, Electronic Music Studios in Britain and Roland in Japan, alongside established musical instrument manufacturers such as Korg and Yamaha. On the musical side, the advent of synths led to the creation of whole new genres. Wendy Carlos' reimaginings of classical compositions were followed by the slicker, less experimental-sounding music of Isao Tomita. His 1974 album *Snowflakes Are Dancing*, based on the work of Claude Debussy, is described by Wikipedia – pretty comprehensively, it has to be admitted – as 'ambient, avant-garde, classical, proto-synthpop, space music'. Much

---

* https://apps.apple.com/gb/app/minimoog-model-d-synthesninetir/ id1339418001

the same description could apply to the work of Tomita's German contemporaries, Tangerine Dream, who also hit the album charts in 1974 with *Phaedra*. Their starting point seems to have been Stockhausen rather than Wendy Carlos – but a very chilled-out, easy-listening kind of Stockhausen.

To describe *Phaedra* and *Snowflakes Are Dancing* as 'ambient', as Wikipedia does, is really something of an anachronism. As a music genre, the term wasn't in common circulation until later in the 1970s. It originated with Brian Eno, who used it to mean music that could be played in the background and didn't require a listener's full attention. That's a very self-effacing thing for a musician to aim at, though not unique. The early 20th-century composer Erik Satie meant much the same thing by the term 'furniture music', which he applied to some of his works. In fact, a lot of the music that's described as 'ambient' is much more substantial than the name suggests – it may be gentle, relaxing and meditative in mood, but that doesn't mean it can't be musically complex and sophisticated.

One of the first works Eno produced in this vein, *Discreet Music* from 1975, has another claim to fame from a technological perspective. It's probably the first example, in popular music anyway, of what might be called 'automated composition'. As Eno explains in the liner notes: 'Since I have always preferred making plans to executing them, I have gravitated towards situations and systems that, once set into operation, could create music with little or no intervention on my part.'

For *Discreet Music*, Eno used a hybrid analogue/digital system – EMS Synthi AKS, which represented the state of the art in 1975. The synthesiser itself was an analogue device, but it was coupled to a digital sequencer, which allowed Eno to programme the notes in advance rather than playing them

in real time. He then added a final stage of analogue processing using a tape-loop feedback system. In what was almost
certainly a first (and last) in the world of pop records, the
back cover of the album includes an engineering-style 'oper-
ational diagram' showing the electronic configuration that
Eno used.

As innovative as artists like Brian Eno, Isao Tomita and
Tangerine Dream were, perhaps the most crucial development in 1970s synthesiser music was made by the German
band Kraftwerk with their 1974 album *Autobahn*. The title
track has a danceable beat, a hummable melody and catchy
lyrics. In short, it's real pop music. Varèse and Stockhausen
had been left far behind, and Kraftwerk were giving the
music-buying public what it really wanted. When a short-
ened version of the title track was released as a single, it
made the top 30 around the world.

Autobahn was half a decade ahead of its time. It's strange,
in hindsight, that it took so long for other musicians to pick
up on Kraftwerk's lead. But eventually the synth-pop wave
hit and shaped the 1980s like no other genre. A new breed of
virtually guitar-less, synthesiser-dominated bands emerged,
from Duran Duran and The Human League to Depeche Mode
and Pet Shop Boys, churning out hit after electronic hit. To
name just a few science-themed ones, there was 'Einstein
a Go-Go' from Landscape, 'Tesla Girls' from Orchestral
Manoeuvres in the Dark and 'She Blinded Me With Science'
from Thomas Dolby. They may be mass-appeal, lowbrow
pop songs, but (for those of us of a certain age, anyway) they
are a cut above the mass-appeal, lowbrow pop songs of any
other decade.

And there were highbrow hits too, like the Art of Noise's
1984 single 'Close (To the Edit)'. With a thumping beat and

Anne Dudley's earworm keyboard riff, it's no surprise that it made the top ten, but the song's real genius lies in the slick editing of multiple textural layers, which channels Edgard Varèse and John Cage as much as Abba or Michael Jackson. It's no coincidence that Anne Dudley has described Cage as 'the patron saint of the Art of Noise'.

Songs like 'Close (To the Edit)' illustrate the way in which sophisticated electronic techniques steadily infiltrated every aspect of popular music, not only through individual instruments like synthesisers, but through the whole process of editing, mixing and production of the finished sound. One member of the Art of Noise, Trevor Horn, went on to make a bigger name for himself as a producer – with artists from Grace Jones and the Pet Shop Boys to John Legend and Robbie Williams – applying what were basically electronic techniques to music that didn't necessarily sound 'electronic' to the listener. In the hands of Horn and his contemporaries, music production became an art form in itself. He reputedly spent three months mixing a single track, 'Two Tribes', from Frankie Goes to Hollywood's 1984 album *Welcome to the Pleasuredome*.

Today, digital production techniques pervade virtually all genres, including ones that aren't usually thought of as 'electronic music', such as hip hop. Back in the 1980s, pioneers of the genre such as Run-DMC and Public Enemy brought electronic 'sampling' – the use of pre-existing audio clips as building blocks of music – into the mainstream, where it's remained ever since. Today, songs by artists like Kendrick Lamar can derive much of their emotional impact from the sophisticated use of sampling. While the samples themselves may be entirely acoustic in nature, the way they're put together is only possible through electronics

– and, ultimately, harks back to 1950s experiments like *Williams Mix* by John Cage and 'Poème Électronique' by Edgard Varèse.

You can also hear echoes of those early pioneers in some of the mellower, more 'ambient' subgenres of electronic dance music. I'm thinking in particular of artists like The Orb and William Orbit, where the overall texture of the sound becomes more important than traditional elements such as harmony and melody – something Stockhausen referred to when he was talking about aleatoric music. When you remove the regular dance beat, as in EDM's spin-off intelligent dance music, the similarity becomes even more obvious. If you took certain tracks by IDM artists like Autechre, Aphex Twin or Venetian Snares and shuffled them with extracts from 'Poème Électronique' and Stockhausen's *Kontakte*, you might have difficulty telling which was which.

If there's one big difference between the electronic music of the 1950s and today, then it lies in the ease with which it can be created. What used to take hours of laborious effort with tape recorders and patch cables can now, thanks to digital technology, be achieved in a matter of seconds with a few mouse clicks. I've been experimenting with this myself since I started writing this book. You may call that playing around when I should have been working, but I call it research. It means I can explain to you in a bit more detail just how simple it is to make music with a DAW app.

The basic function of the DAW is to combine a number of simultaneous audio tracks, possibly dozens of them, into a coherent song that can then be exported in a standard format such as MP3. The input for a track could itself be an MP3-style audio clip – anything from a recorded vocal or instrumental to a sound-effect sample – but there's another option too. A term I've used several times in this book is MIDI, which stands for Musical Instrument Digital Interface. It's the standard protocol for communicating musical information, as opposed to an audio signal. If an .mp3 file can be thought of as the digital equivalent of a vinyl record, then a .mid file is the equivalent of a printed musical score. It doesn't contain any audio information, just instructions as to which notes to play. This means you can put MIDI files into a DAW track, but you then have to feed them into a synthesiser plugin before you hear any sound.*

This is where the fun really starts, because synth plugins produce such a wide range of weird and wonderful sounds that you can quickly turn quite a banal little tune into surprisingly sophisticated-sounding music. If you add in your pre-recorded audio clips – which you can loop or stutter or speed up or slow down or modify in dozens of different ways – and use a drum-machine plugin to provide a rhythm track, then you're pretty much there. You've created a song with a much slicker, more professional sound than any amateur musician could have come up with 30 years ago.**

---

* Since I understand music notation, I use a score editor to create MIDI files, but they can also be produced with piano-roll editors or MIDI keyboards.

** I've put a few of my own musical experiments, in a range of genres, on YouTube: https://www.youtube.com/playlist?list=PLvQfFzOIn EpTjvazcXJE9ykgCuxzatY_v. This gives an idea of what an aspiring beginner can produce after a few months of playing around. At the other extreme, here's an astonishing video from a professional producer who manages to create the basic material for five different EDM songs in the space of 25 minutes, see: 'Making 5 EDM Tracks In 25 Minutes – FL Studio 20 Tutorial', Arcade, YouTube, https://www.youtube.com/watch?v=rhX63VJXd34

The ease with which new music can now be created means that larger quantities of it are being produced than ever before, with more new songs released each week than a person could comfortably listen to in a year. In what I still insist on thinking of as the 'electronic music genre', singular, the AllMusic site lists no fewer than 75 subgenres – all the way from experimental ambient and trip-hop to techno-dub and progressive trance.

The fact is that, for the first time in history, anyone with the talent and motivation can make a professional-quality recording. There's no need for a studio, or expensive equipment or a recording contract or the right sort of contacts. Back in Chapter 1, I described how Jacob Collier makes all his music using DAW software on an ordinary desktop computer. What I failed to mention is that he does it in his bedroom. His first album, in 2016, was actually called *In My Room* – but all his later albums could equally well have had the same title.

These days, there's no reason a bedroom-produced album can't fly as high as a studio-produced one. In 2019, Billie Eilish's first album, *When We All Fall Asleep, Where Do We Go?*, reached the number one spot in the United States, the United Kingdom and many other countries. Yet it was recorded and produced by her brother Finneas in his bedroom, using the same Logic Pro software that Jacob Collier uses. Even Kanye West's quadruple-platinum debut *The College Dropout* (2004) was largely a bedroom production – albeit using a dedicated standalone workstation called the Roland VS-1680. You can pick up a used one of those on eBay for under £200.

If you don't like the idea of parting with money (who does?), you still have options. If you already own a Mac computer, you'll have Logic Pro's little brother, GarageBand, already installed on it – and that's a perfectly adequate DAW to get started with. Even better, and regardless of what brand of computer you have, there's Waveform Free – which is the program I've been using. And it really is free, not just a time-limited or functionality-reduced demo. Whatever DAW you acquire, you have access to the same enormous range of plugins, many of them free, which allow you to emulate all sorts of synthesisers, drum machines, samplers, filters, equalisers and a whole range of effects, from distortion and acoustic 'glitches' to echo and reverb.

With those plugins, you can do anything that the greats of electronic music, from Stockhausen and Delia Derbyshire to Wendy Carlos and Kraftwerk, could. More, in fact – though some of the plugins in the 'more' category are of questionable artistic value. There's the notorious 'autotune', for example, which allows you to pull out-of-tune vocals back into tune. But on occasion, even this has been adapted to more creative purposes – for example, in the Radiohead song 'Pulk/ Pull Revolving Doors' – where it's used to convert spoken dialogue into pseudo-singing, and arguably at least one genuine pop masterpiece: Kanye West's 'Love Lockdown', which employs autotune to create searingly powerful vocals that couldn't have been achieved any other way.

In a similar category is 'quantisation', which was briefly referred to in Chapter 1. Wikipedia's description of this, intentionally or not, reads hilariously:

> In digital music processing technology, quantisation is the studio-software process of transforming performed musical notes, which may have some imprecision due to expressive performance, to an underlying musical representation that

eliminates the imprecision. The process results in notes
being set on beats and on exact fractions of beats.*

If you read that carefully, it implies that quantisation is
designed to *eliminate* expressive performance from recorded
music. Just as autotune pulls out-of-tune notes back to the
'correct' pitch, so quantisation pulls notes that are slightly
early or late back to the 'correct' beat – or half-beat or
quarter-beat or whatever. But what great musician doesn't
sometimes play notes slightly early or late, for deliberate
emotional effect? Fortunately, a good producer will use quantisation very selectively to correct wrong-sounding notes, not
to ruin deliberately mistimed ones.

For better or worse, there are countless ways that digital
technology can aid music production. But it goes even further
than that. As soon as we start to think of an audio waveform
as 'data', there's no end to the ways we can generate or
manipulate it. It's time to revisit a subject we discussed earlier, musical algorithms – but this time we'll look at it from
a thoroughly digital perspective.

Algorithms revisited

To a computer, as I said earlier, all data looks the same. It
doesn't matter if it's a symphonic masterpiece or the fluctu-
ating electromagnetic emissions from a distant galaxy, it's
just a long string of numbers to a computer. And that raises
an interesting question: what would those emissions sound
like if we asked a computer to treat them like an audio signal?

* https://en.wikipedia.org/wiki/Quantization_(music)

As we saw in Chapter 2, human hearing can detect frequencies between about 20 Hz and 20 kHz. Many naturally occurring oscillations fall well outside this range. Ultra-low-frequency waves in the Earth's magnetosphere can be measured in millihertz, or thousandths of a cycle per second, while the radio waves emitted by a collapsing star fall in the gigahertz range – billions of cycles per second. That's not such a problem as it might appear, though, because the frequencies can be scaled up or down to bring them into the audible range.

This process has become so well-established in recent years that it's been given a formal name: astronomical sonification. From an aesthetic point of view, this may seem a fairly haphazard way to create music, rather like a digital-age version of John Cage's *Atlas Eclipticalis* that was mentioned in the previous chapter. But it has practical uses for astronomers too, as it is a useful way to analyse data alongside more traditional visualisation techniques.

Even from a layperson's point of view, astronomical sonifications can have a surprising appeal, often sounding as spookily alien as their extraterrestrial origin might lead you to expect. One of the most widely heard examples came from the European Space Agency's Rosetta mission to comet 67P in 2014. Called the 'Singing Comet', this particular sonification was based on low-frequency oscillations in the comet's magnetic field, which were scaled up by a factor of 10,000 to make them audible.* The result was one of the first pure audio clips to go viral, with almost 6 million listens on SoundCloud and mass coverage in the world's news media.

---

* 'The "Singing Comet" 67p/Churyumov-Gerasimenko', European Space Agency, https://sci.esa.int/web/rosetta/-/55034-the-singing-comet-67pchuryumov-gerasimenko

Perhaps the most obvious source of outer space sounds, given the use of radio for sound broadcasting on Earth, is radio astronomy. One of the pioneers of 'acoustic astronomy' in this context was Fiorella Terenzi, who had the idea of converting radio emissions from distant galaxies into audio form while working on her doctoral thesis at the University of California in 1987. Now a professor at Florida International University, Terenzi has applied similar techniques – initially as an aid to data analysis, and more recently as a form of science outreach – to other astronomical sources ranging from the planets Jupiter and Saturn to distant pulsars, quasars and X-ray binaries.

Terenzi's sonifications, like Rosetta's 'Singing Comet', stay pretty close to the raw astronomical data, with the bare minimum of processing to make them audible to our ears. Other people, however, have gone a step further to produce works that could almost be called 'astronomical music'. In 2011, for example, astronomers Alex Parker and Melissa Graham created a 'Supernova Sonata' using data from 241 supernova events. All the musical elements, such as pitch, volume and instrumentation, were derived from different astrophysical parameters. The following year, NASA itself got in on the act, when Sylvia Zhu of the Goddard Spaceflight Centre turned data from a gamma-ray burst into something approximating music. To achieve this, she had to scale down each gamma-ray photon from its original, enormously high, frequency into the audible range, as well as slowing down the arrival rate of photons to a sensible tempo.

Even a few professional musicians have made use of astronomical data. Thomas Dolby, for example, incorporated some of Fiorella Terenzi's sonifications on his 1994 album *The Gate to the Mind's Eye*, while the reggae band Echo

Movement used data from the Kepler space telescope on *Love and the Human Outreach* (2012). But perhaps the most unlikely practitioner of astronomical music is former Grateful Dead drummer Mickey Hart, who teamed up in 2012 with the cosmologist George Smoot from the Lawrence Berkeley Laboratory in California.

A few years previously, Smoot had won the Nobel Prize for his work on the cosmic microwave background (CMB) – the oldest radiation in the universe, which Hart has described as 'beat one'. The collaboration resulted in Hart's album *Mysterium Tremendum* as well as a twenty-minute audio-visual experience called *Rhythms of the Universe*, first shown at the Smithsonian Air and Space Museum in Washington, DC. Taking Smoot's data on the CMB as a starting point, Hart worked with Lawrence Lab scientists to combine it with other astronomical data and transform it into something that actually sounds like music – a feat he repeated in 2018 with a second astronomically inspired event called *Musica Universalis* at the Hayden Planetarium in New York.

If you're attracted to the idea of composing music where all the notes are handed to you on a plate, but you don't have a radio telescope or other source of astronomical data to hand, there are other options available. As with so many things these days, there's free software you can use. In this case, it's a specially designed 'algorithmic composition' tool called OpenMusic, created by the Institute for Research and Coordination in Acoustics and Music in Paris.* As a long-time computer geek, I find OpenMusic very easy and intuitive to use, although a traditional musician might

* https://support.ircam.fr/docs/om/om6-manual/co/OM-Documentation.html

disagree – but then again, a traditional musician isn't going to need it.

OpenMusic comes with a number of built-in libraries that you can use to generate music. There's one library called OMChaos, for example. That may not sound very promising, but it doesn't refer to the colloquial use of 'chaos' but to mathematical 'chaos theory', which you may have seen used to create attractive visual patterns like the Mandelbrot set. The same equations can be used to produce musical patterns too.

Another OpenMusic library is OMAlea, designed to assist with aleatoric composition. As you'll recall from the last chapter, this comes from the Latin word for dice, and basically means music with a random element thrown in. The library contains various procedures taken from the field of mathematical statistics, including a particularly useful one called a Markov chain. It's worth looking at how this works in a little more detail.

Let's start by assuming that we want to write a money-spinning, smash-hit tune but lack the necessary musical talent (I'm speaking mainly for myself here). So we're going to get our computer to do it for us. Reducing the problem to its simplest level, we want the computer to string a long series of notes together to make a melody. We know from earlier chapters that there are twelve different pitch classes, so we could start by asking the computer to pick a selection of these at random. But if we do that, we're likely to find the result disappointing. At best, it's going to sound like Schoenberg on a bad day, which isn't the effect we want at all. The aim, remember, is to sell lots of records.

Taking a slightly different tack, let's pick a song we can use as a model – 'Shape of You' by Ed Sheeran, for example.

Don't worry, we're not going to copy it, just do a little statistical analysis on it. As it happens, the song only contains five different PCs: C♯, E, F♯, G♯ and B. A music-theory pedant might jump up and down at this point and say: 'Aha, C♯ minor pentatonic', but we don't really care about that. We've got our computer programmer hats on, and it's all just data to us.

So, we could refine our instructions to the computer and ask it to pick a random sequence of notes, just drawing on those five PCs with equal probability of occurring. Even better, we could look at how often the different pitches appear in Sheeran's song, and tailor the relative probabilities accordingly. As it happens, F♯ is the commonest PC, occurring around a third of the time, followed by E a quarter of the time and then the other notes less often.* So, we could make sure they appear with the same frequencies in our own song.

The result probably wouldn't sound too bad, particularly if we added a funky rhythm and a suitable chord progression as accompaniment. But as I said in the previous chapter, musicians have very subtle rules regarding the role that each note plays in a scale, and if we picked them at random they wouldn't obey those rules. As Eric Morecambe said of his mangled rendition of Grieg's Piano Concerto in 1971, it would be 'all the right notes, but not necessarily in the right order'.

This is where Russian mathematician Andrei Markov comes in. He found that, in many random processes that occur in nature, the probability of a particular state arising depends on the state immediately preceding it – and he

---

* I'm being deliberately vague here, to stay on the safe side of the 'fair use' copyright rules.

developed a mathematical formalism to handle such situations. What OpenMusic's OMAlea library does is apply the same formalism to musical notes. The probability of a particular note arising depends on which note precedes it.

Let's go back and have another look at 'Shape of You'. We already know that F♯ is the commonest PC, so we'll start with that. According to OMAlea, there's a 35 per cent chance that F♯ is followed by E, 30 per cent that it's followed by another F♯, 24 per cent by G♯ and 11 per cent by C♯. It's never followed by a B. So we can use those probabilities to choose the second note of our song. The most likely option is E, and if we pick that we'll have to go back to OMAlea to find out the probabilities for the next note, which are going to be different moving from E than they were moving from F♯.

The result, if we carry on like this, is called a 'first-order Markov chain' because each note only depends on the single note preceding it. If we make matters more complicated by looking at two consecutive notes instead of just one, that would be a second-order chain, and so on. And these days, there are much more sophisticated extrapolation methods than Markov chains. Another OpenMusic library, called LZ, implements the Lempel–Ziv algorithm, which works on rhythm and chords as well as melody.

The problem with the LZ algorithm, and with higher-order Markov chains for that matter, is that it's so good at its job that the output ends up sounding actionably similar to the input. But there's no reason why the input has to be just one song – it could be every hit single from the past twenty years. You'd need to do some pre-processing to put them all in the same key and tempo, and you'd probably have to run the program a few hundred times before it produced something worth listening to – but eventually it would. It might not be the money-spinning hit you were hoping for, but it would be enough to impress friends and family. As an example, you can see what I made from it on my YouTube channel.* I think I've scrambled the melody enough that Ed Sheeran's original doesn't come through, but if it does, I could throw his own words back at him: 'There's only so many notes and very few chords used in pop music.'

If you stop and think for a few seconds, this is getting a little scary. I'm talking about things that can be done now, by a complete amateur, using open-source software that can be downloaded for free. If that's where the world is now, how long will it be before a computer can write music as well as a human? And why is that so much more frightening than the idea of a computer designing a car, or writing an online encyclopaedia entry? What is it about humans and music anyway that makes it such an emotive topic for all of us? These are just some of the questions we'll look at in the final chapter.

---

* https://www.youtube.com/watch?v=3qLn9HqN79k

MUSIC AND THE BRAIN 5

There's something strangely paradoxical about music. As we've seen so far, music basically consists of sound waves whose behaviour can be fully explained by the well-established laws of physics. On that basis, an alien arriving from a music-less planet could be forgiven for thinking that it is a purely objective phenomenon, and that all human beings perceive it in the same way. But of course, that's not the case. Music is one of the most deeply emotional of all forms of art, to which everyone has their own individual response. So where does this baffling paradox – between the physical objectivity of sound waves and the emotional subjectivity of music – come from?

The answer lies inside the human brain. It's an enormously complex system, many different parts of which are involved in the creation and perception of music. The basic 'input data' – the sound wave – enters our brain through the auditory cortex, which is directly connected to our ears. This processes the sound in a way that is actually quite machine-like, analysing it into its component frequencies.

Our perception of the sound as 'music' only starts when it's passed to the prefrontal cortex, which responds in terms of our conscious expectations, and the extent to which these are satisfied or subverted by what we're actually hearing. These reactions are intimately tied to our musical memory, which resides in another brain area called the hippocampus. Finally, our 'gut-level' emotional response to the music comes, not literally from the gut, but from other areas of the brain called the amygdala and nucleus accumbens.

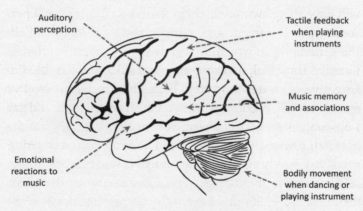

Auditory perception

Tactile feedback when playing instruments

Music memory and associations

Emotional reactions to music

Bodily movement when dancing or playing instrument

Various areas of the brain are involved in our response to music.

The psychology of music

To anyone other than a scientist, this emotional response to music is really the only element that matters. And for most people it can be a deeply personal thing. In researching this book, I've frequently encountered user reviews on sites like AllMusic, or comments on YouTube, that describe some random album – often one that I personally find boring or trite – as 'the greatest music of all time'. They're presenting what's really just a subjective judgement as if it's a black-and-white statement of fact. As Elizabeth Hellmuth Margulis says in her 'Very Short Introduction' to the psychology of music, it can be 'difficult for people to imagine that their own fundamental responses to sound are not universally shared'.

The fact is that, even within the relatively narrow confines of Western music, there's a huge range of different genres and styles. It is therefore inevitable that individuals will have their own preferences, based on a mixture of personality, environment and, perhaps most important of all, the generation in which they grew up. Despite my attempt to make this book as genre-encompassing as possible, I'm sure my own tastes have crept through in the examples I've chosen to talk about. Applying a bit of self-analysis, I'd say I probably respond most positively to classical music of the first half of the 19th and second half of the 20th centuries, progressive rock of the 1960s and 1970s, synth-pop of the 1980s and various ambient, New Age and minimalist music genres of the 1990s and later.

I'm not sure that there's much of a pattern there, but if there is a linking factor, it probably has more to do with what musicians call timbre – i.e. instrumental and vocal textures – than with the more commonly analysed elements of music such as rhythm and harmony. As Daniel Levitin points out in his book *This is Your Brain on Music*, timbre seems to hold a 'privileged position' for the majority of people, placing it above all other aspects of music in the listening experience. So it's often timbre alone that determines a person's musical taste. Think back to the 'heavy rock' arrangement of Schubert's 'Erlkönig' that I mentioned a couple of chapters ago.* Since I created this myself, I can assure you that the pitches and durations of the notes are almost exactly as Schubert wrote them, and it's only the timbral context that has been changed. Yet many Schubert aficionados are going to shudder at the very sound of it, while some rock fans – even those who aren't overly keen on early 19th-century art songs – might actually quite like it.

Turning from timbre to the more traditional musical elements of rhythm, melody and harmony, it also seems to be the case that different people – even when listening to the same piece of music – may focus their attention on different elements of it. So an EDM fan might be unimpressed by a classical string quartet because there's no discernible beat, while a devotee of chamber music might get bored with a dance track because it goes on and on without changing key. There's no right or wrong here – it's just the different perspectives people have. For example, the Hot Science series editor, Brian Clegg, who happens to sing bass in a choir, tells me that what makes a piece for him is primarily its harmonies. For his wife, who sings soprano in the choir, it's the tune. One explanation for that might simply be their different roles in the choir, with the soprano normally carrying the melody and the bass underpinning the harmony.

Another reason why different people respond to the same piece of music in different ways depends on how familiar they are with that particular musical style. Inevitably, people of different generations and different cultural backgrounds will have different ideas of what 'good' music sounds like. So, as we get older, our openness to new musical experiences

* https://www.youtube.com/watch?v=u3nvuNC19vk

depends on how familiar they are in the context of what we already know.

This can be illustrated in a scientific-looking way using something known as the Wundt curve. I say 'scientific-looking' rather than 'scientific', because the graph makes no attempt to put numerical values on the axes. It's just a hand-waving way of saying that, in the case of many psychological parameters, 'as the stimulus increases the response goes up, hits a peak at some sweet spot, then drops down again'. The curve was originally described in 1874 by Wilhelm Wundt, who was the first person to self-identify as a psychologist.*

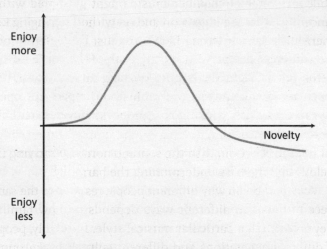

The Wundt curve shows how our reaction to new music depends on its novelty.

---

* So, depending on your viewpoint, either one of history's great pioneers or one of its great villains.

My diagram shows one of several ways that the Wundt curve can be applied to music perception. Suppose we're presented with a new piece of music that we've never heard before. If it's closely similar in form and style to music that we're already familiar with, then our reaction is likely to be pretty neutral. The more different it is, up to a point, the more we'll sit up and take notice. At some point it will hit a sweet spot where it's just novel enough to be really attention-grabbing, but not different enough to be alienating. After that our enjoyment may decline and eventually turn into active dislike.

While virtually everyone will follow something akin to a Wundt curve when it comes to listening to a new piece of music, its parameters will vary from person to person. For one thing, what constitutes 'familiarity' will depend on an individual's age and prior listening habits. Even allowing for that, different people will put the peak of the curve further to the left or right depending on how adventurous their tastes are. You could say the same about a person's openness to trying new cooking recipes or holiday destinations, for that matter.

So, musical taste, like any other personal preference, is highly subjective. But is there any sense in which music can be measured on an absolute scale, that transcends personal and cultural differences? Many people seem to believe this is the case, from those annoying AllMusic reviewers and YouTube commenters to at least some musicians themselves. Here's Beethoven's rather grandiose take on the subject, as recorded by Bettina von Arnim in 1810:

> Music is the one incorporeal entrance into the higher world
> of knowledge which comprehends mankind but which

mankind cannot comprehend ... Every real creation of art is independent, more powerful than the artist himself and returns to the divine through its manifestation. It is one with Man only in this, that it bears testimony of the mediation of the divine in him.

He seems to be saying that music has a significance that goes beyond the merely human world. That sounds pretty far-fetched, given that music is created by human minds for human ears. Yet I've come across a couple of semi-persuasive arguments that suggest how Beethoven's view might actually be correct. It was a few years ago, when I was writing a book called *The Science of Sci-Fi Music*. The great thing about science fiction – I mean the serious, intellectual kind, not mass-market blockbusters – is that authors can try to get inside non-human minds and work out how they might perceive things such as music.

One of the most thoroughly alien of all science fictional extraterrestrials has to be the 'Black Cloud', in Fred Hoyle's brilliant 1957 novel of that title. It's a vast cloud of interstellar gas made of molecules that have organised themselves into a gigantic brain, much older and far more intelligent than human beings – with whom it learns to communicate via radio waves. Yet when a piece of music (by Beethoven, as it happens) is transmitted to it, the cloud is impressed by it. That may seem improbable for an entity with no conception of acoustic sound or human emotions, but to the cloud the music is simply a satisfying pattern of rhythmic oscillations. As one of Hoyle's characters remarks:

Our appreciation of music has really nothing to do with sound, although I know that at first sight it seems otherwise. What we appreciate in the brain are electrical signals that we receive from the ears. Our use of sound is simply a convenient device for generating certain patterns of electrical activity. There is indeed a good deal of evidence that musical rhythms reflect the main electrical rhythms that occur in the brain.

Another argument that impressed me came from musician and science-fiction author Langdon Jones, in a story called 'The Music Makers' dating from 1965. It's set on the planet Mars, where two characters get into a debate as to whether human music, such as Alban Berg's Violin Concerto, would have had any meaning for the long-dead Martian natives. One character takes the view that music is much too human-centric, while the other – reflecting what I suspect was Jones' own view – argues that it's only our *response* to music that's parochially human, while the music itself is universal:

> Great music is something absolute, to which we respond with human emotions. What Berg created was something like these dunes, permanent and absolute, and what feelings Berg creates in us are in essence produced no differently than those caused by the dunes.

The reality is that arguments like these, pro and con, can go on forever – because no one really understands how music affects us in the ways that it does. The same is true on the creative side – ultimately, no one knows where music comes from in the first place. There are countless anecdotes of whole works appearing, fully formed and unbidden, in a composer's head. One of the most amusing is Frank Zappa's description of the origin of 'Who Are the Brain Police?', a song he wrote in 1966: 'At five o'clock in the morning someone kept singing this in my mind and made me write it down.'

It's pretty obvious from his jokey tone that Zappa realised the music came from his own subconscious. But to someone who was less of a cynical realist, it may have seemed genuinely to have emanated from an outside source. One musician who often talked in such terms was Stockhausen, who ascribed many of his works to 'cosmic inspiration'. As he well knew, he was far from the only one:

> I know that Bach, Beethoven, Brahms, Stravinsky and a few
> other composers of the past have recognised the supremacy
> of intuition, based on the composer being a medium. He
> is a mouthpiece of the divine.

Stockhausen said this in an interview with Robin Maconie, and it's only fair to add that, later in the same interview, he acknowledged that 'a memory stored in the psyche is sometimes taken for a cosmic influence'.

Perhaps the most astonishing example of musical intuition is the work of Rosemary Brown. Between 1964 and her death in 2001, she produced a whole stream of works that she said had been 'dictated' to her by the spirits of deceased composers from Beethoven, Liszt and George Gershwin to – following their deaths – Stravinsky and John Lennon. Unlike many self-professed psychic mediums, no one who has studied her case seriously suspects Brown of conscious fraud. There seems little doubt that she really did perceive the music as originating the way she described it, as coming from a source outside her head. Given the mysterious nature of musical creativity, that's really not that hard to believe.

What's really remarkable, however, is that not all her compositions are simple pastiches of the style of one composer or another. In some cases, we're talking about composers – Beethoven, Liszt and Stravinsky in particular – whose styles continued to evolve throughout their lives. Reproducing what they might create 'post-mortem' isn't simply a matter of copying their best-known works, which were generally produced decades before their death. Liszt's final works, for example, seem to foreshadow the next generation of composers, such as Scriabin and Debussy. And that's something we can hear in one of Rosemary Brown's most remarkable pieces, 'Grübelei' ('Brooding'), which was supposedly transmitted by Liszt in 1969.

The thing is that 'Grübelei' – the creation of which was witnessed by a BBC television crew – doesn't sound like anything Liszt actually wrote. Instead, as the composer Humphrey Searle observed: 'It could well be something that he would have written had he lived another two years.' And Searle was as qualified as anyone to say that. If you've ever bought a CD with Liszt's music on it, you'll notice that all the works have catalogue numbers prefixed with the letter S. That S stands for 'Searle', because he was the one who created the catalogue.

Who knows where Rosemary Brown's music actually came from? Presumably it was from somewhere inside her brain, but – as we saw earlier – the human brain is an enormously complex system, many parts of which are actively involved in the creation of music. One way to find out what happens during the process of music creation is via a magnetic resonance imaging (MRI) scan – and specifically through the rather oddly named technique of 'functional MRI'. That's not to distinguish it from non-functional (i.e. broken) MRI, but because it can identify which areas of the brain are functioning during a particular activity. It's been used to analyse brain activity in jazz musicians while they are improvising on a keyboard. Perhaps the most striking finding relates not to the brain areas that are active during improvisation, but the ones that aren't. In particular, the part of the brain that's responsible for self-monitoring, called the 'dorsolateral prefrontal cortex', shuts down completely. So, in a sense, all those people from Beethoven and Stockhausen to Frank Zappa and Rosemary Brown were right – music does come from 'outside' their own persona.

Although psychological studies of music generally focus on its emotional impact, it can have other effects on our brains as well. One phenomenon that has received a lot of attention in recent years is 'autonomous sensory meridian response', or ASMR. This is the weird effect whereby hearing certain sounds – not musical ones, but sounds like soft tapping or whispering – can make you feel as if your skin is being tickled. While most people seem to be susceptible to ASMR to some degree, it's really a low-level form of a much rarer condition called synaesthesia, whereby one type of sensory input is perceived as a completely different sensation.

In one form of full-blown synaesthesia, musical sounds can be perceived as colours. Although this is technically a disorder, you can see how it might be beneficial in a musical context. In fact, many musicians who have 'suffered' from synaesthesia – from Liszt and Messiaen to Billie Eilish – actually view it as a gift. The Russian composer Alexander Scriabin even went so far as to create a synaesthesia-emulating instrument called the clavier à lumières. Essentially a standard keyboard that controls coloured lights instead of musical notes, it was used at a performance of Scriabin's orchestral work *Prometheus: Poem of Fire* in New York in 1915.

The clavier à lumières never caught on, but ASMR has become a worldwide phenomenon, with something like 15 million YouTube videos devoted to it. It's most commonly billed as a gentle cure for insomnia, which has roots in one of the oldest physiological uses of music – as a lullaby to get children to sleep. And adults, too, can be sent to sleep by the sound of music. There's a great anecdote in this context about the most famous lullaby-writer of all, Johannes Brahms, and the composer with whom he's inextricably linked in many people's minds, Franz Liszt.* I'd like to say that Liszt – who was the older of the two by more than two decades – fell asleep while Brahms played his lullaby, but it was actually the other way around. The twenty-year-old Brahms was the one who fell asleep, while Liszt played his recently completed Sonata in B minor (S.178). Contrary to some accounts, Liszt doesn't seem to have been overly offended by this – he knew the younger musician was exhausted after long journey.

Another person who was supposedly sent to sleep by a piece of music was Count Herman Karl von Kayserling, the Russian ambassador to Saxony in the 1730s. A sufferer from chronic insomnia, he asked the composer J.S. Bach to produce some music to help him sleep. The result was the 80-minute keyboard piece now known as the *Goldberg Variations*, after the harpsichord player who first performed it for Kayserling. Anyway, it seems to have done the trick, and Kayserling was soon fast asleep.

---

* In certain circles, 'Brahms and Liszt' is rhyming slang for 'thoroughly inebriated'.

Fast forward to the 1970s, when a study by the Bulgarian scientist Georgi Lozanov indicated that certain sections of the *Goldberg Variations* have the effect of slowing down various bodily processes. This wasn't unique to Bach; much the same effect could be produced by the slow movements of several other 18th-century keyboard pieces. Rather than recommending them as cures for insomnia, however, Lozanov promoted them as aids to meditation – something that was becoming very fashionable at the time.

Although it created a minor fad in New Age circles, the 'Goldberg effect', as it might be termed, never made mainstream headlines. Twenty years later, however, something very similar did. This was the notorious 'Mozart effect', whereby – according to a rather shoddily conducted 1993 study – listening to Mozart can make a person smarter. Or rather, that's the way the study was often summarised in the media. What the researchers actually claimed was that their experimental subjects showed significantly better performance in spatial reasoning tasks if they listened to Mozart's music – as opposed to doing nothing – immediately before attempting the task.

The results weren't faked or flawed, and the conclusion, stated in the way that I just did, was correct enough. But the study was still hopelessly misleading. It was a case of the researchers asking a silly question and getting a silly answer. All they asked was whether listening to Mozart helps a person to perform certain mentally demanding tasks better than if their brain had just been idling. But why Mozart, specifically? Later studies repeated the experiment but substituted other types of music – such as Ravi Shankar or The Beatles – and found exactly the same effect. It doesn't even need to involve music, for that matter; any kind of mental stimulation will do the trick. So the 'Mozart effect' was well and truly debunked – but not until record companies fortunate enough to have Mozart CDs in their catalogues had made a tidy profit from it.

The fiasco of the Mozart effect brings home a major disadvantage of traditional psychology studies: researchers can only address very specific questions that can be examined under laboratory conditions – such as the difference between listening to Mozart and doing nothing. It would be much more helpful if we could encompass all of the possible activities a person might be engaged in. These days, there's a way we can get much closer to that – and it doesn't involve psychology laboratories at all.

Music and big data

As you already know, the big tech companies are spying on your every move. More specifically, they're keeping a record of everything you do with the aid of digital technology – though, to be fair to them, usually in an anonymised way. So they may record the fact that *someone* did an embarrassingly slow jog around the park this morning, without attaching their particular name to it. In the context of music, streaming services like Spotify and Apple Music collect vast amounts of data on people's listening habits.

Clearly this tells the companies, and any researchers they share the data with, what the current most popular songs are – and with much greater accuracy than old-style record sales or radio audience figures. But they can go much further than that. By looking at other data recorded from the same device, they can correlate music listening with a host of other factors, from the geographical location and time of day to a person's age and social connections. It's also possible to see if the listener was involved in any other digital activity at the time, and possibly – if they are wearing a fitness tracker – physical activity too.

It's a scary prospect for most of us because no one likes the thought that their every move is being watched and analysed, even if the data remains anonymous. But it does have its benefits. If you listen to a familiar song on a music streaming service, after it finishes it may segue into another track that the software 'thinks' you might like. It may be something you've never heard before, and on occasion it can turn up an unexpected gem. The same 'big data' approach is also invaluable to psychologists of music, since it has a much firmer basis in reality than the kind of laboratory studies that produced the illusory and misleading Mozart effect. The same data can also be used by the music industry itself, to help shape the sound of new music. As we'll see shortly, this can have both positive and negative effects on the creative process.

Digital data – the total amount of which is currently approaching 100 trillion gigabytes – isn't limited to newly acquired data. People are steadily digitising all the material from pre-digital times, which in the musical context means everything from medieval manuscripts to piano rolls and wax cylinder recordings. This makes it much quicker and easier to analyse the material, both from a musical and scientific point of view. It's now possible to carry out 'corpus studies' that take into account every single piece of music in a particular genre or category and look for trends, similarities and differences.

As an example of a corpus study, Elizabeth Margulis refers to a phenomenon we encountered in Chapter 3 – the rise and fall of the diminished seventh chord in classical music. That was the one Schoenberg was so scornful of in 1911, because of the way it was used to represent more or less any emotional extreme. Yet a century earlier, when it was still a rare and surprising occurrence, it had a very different reputation. The corpus study allowed researchers to trace the increasing occurrence of the chord through the 19th century and its subsequent decline in the 20th century, and correlate this with favourable and unfavourable comments about it from academics and the music press.

Corpus studies of this kind are far more useful to scientists than they are to musicians. A musician will take one look at the conclusion – 'diminished sevenths were a 19th century fad that had all but disappeared by the mid-20th century' – and simply see it as stating an obvious, well-established fact. But scientists really don't like anecdotal or hand-waving statements, no matter how true they may be, because there's nothing they can do with them. They want numbers they can apply various statistical methods to, and that's what corpus studies give them.

Another scientific study which, at first sight, might appear to be stating the obvious relates to the evolution of popular music in the half-century between 1960 and 2010.* It consists of a massive statistical analysis of the Billboard Hot 100 throughout that time, and shows, for example, how hip hop grew from almost nothing prior to the mid-1980s to dominate the post-2000 charts. 'Easy listening' music, which was a big thing in 1960, has steadily declined in popularity ever since then. Female vocalists are a major feature at both

---

* Matthia Mauch et al., 'The evolution of popular music: USA 1960–2010', *Royal Society Open Science*, vol. 2, 5 (May 2015), https://royalsocietypublishing.org/doi/10.1098/rsos.150081

ends of the survey but suffered a dip in popularity from the 1970s to the 1990s. The opposite is true of electronic music, which saw its peak in those decades, but is now heading back to its 1960s level. And so on.

OK, so you probably knew most of that already. But the study puts hard numbers on it, on a week-by-week basis, in a way that would have been impossibly time-consuming in the pre-digital age. This paves the way to statistical analyses that can genuinely tell us something new – as happened in 2018 when one was used to shed light on a long-standing pop mystery.

In the early days of The Beatles, John Lennon and Paul McCartney agreed to share writing credits on all their songs – although more often than not they were actually the work of just one or other of them. This could have led to no end of confusion, but fortunately for music history the two were usually in complete agreement as to which of them actually wrote a particular 'Lennon and McCartney' song. But there's an exception in the case of 'In My Life', from the 1965 album *Rubber Soul*. While everyone agrees that Lennon wrote the lyrics, both he and McCartney claimed to have written the melody.

So where does that 2018 statistical analysis come in? A group of researchers – including Jason Brown, the Canadian professor who helped us identify the opening chord of 'A Hard Day's Night' in Chapter 2 – analysed the note sequences and chord progressions in 70 Lennon–McCartney songs from 1962 to 1966. They compared these with the ones occurring in 'In My Life' and concluded that, from a statistical point of view, the chance that Paul McCartney had written the music was less than one in 50. So, a posthumous win for John Lennon – but not really bad news for Paul either. After all, they do say that if you can remember what happened in the 1960s, you can't have been there.

Another statistical study that caught my eye relates to the use of different keys in 18th- and 19th-century music. I mentioned earlier how I've always been mystified as to why a composer chose to write a particular work in one key and not another, and then considered the key important enough to be stated explicitly in the work's title. The study I'm talking about – actually a student project – looked at a large body of works by a range of composers, and plotted charts of the frequency of occurrence of different keys.*

Despite the fact that keys are such a mystery to me, I do tend to notice and remember them in the titles of works by certain composers, such as Haydn, Mozart, Beethoven, Schubert and Mendelssohn, where they're often the main identifying element. One thing I spotted fairly recently, for example, is that a lot of Haydn's most striking symphonies and string quartets are in the key of D. So I was interested to see from the statistical data that D was indeed Haydn's most frequently used key. There are a couple of pragmatic ways you could explain that. Maybe it was simply the fashion of the time and place where he worked that audiences liked D, or maybe the orchestras and instrumental ensembles he wrote for found it easier to play in D.

These are both great theories, but they won't wash. Mozart worked at exactly the same time and in the same part of the world as Haydn, yet his chart doesn't show the same peak for D. With Mozart, D is on the same level as two other

---

* Ethan Paik Marzban and Caren Marzban, 'On the Usage of Musical Keys: A Descriptive Statistical Perspective', *The Journal of Experimental Secondary Science*, vol. 3, 3 (2014), http://staff.washington.edu/marzban/music.pdf

keys, F and B♭, which Haydn seems less enamoured with. In fact, Haydn doesn't seem to have been very fond of F at all.

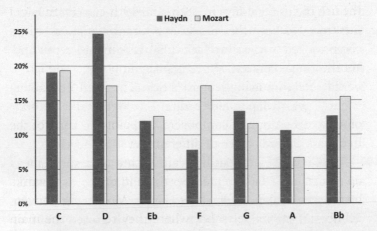

Although they were contemporaries who wrote for the same instrumental combinations, Haydn and Mozart had subtly different tastes in keys.

So it looks like the choice of key is down to personal preference, placing it in the category of 'musical psychology' that we started this chapter with. It seems that, in the period we're talking about, both musicians and listeners tended to associate different keys with different moods or emotions. For example, one account from around 1800 describes D as the key of 'triumph and rejoicing', F as 'friendliness and calm' and B♭ as 'cheerfulness and hope'. While this might sound like mumbo-jumbo on a par with an astrological horoscope, if you think about it for a moment, you'll see how it could easily become a self-fulfilling prophecy. If composers believed in these associations, they'd put them in their music, wouldn't they?

These days, thankfully, people don't worry too much about what key a piece of music is in. If the harmonic structure of a pop song carries any emotional weight, then it's through the use of different chords rather than specific keys. This, too, has been the subject of a recent statistical study, which looked at the correlation between chord types and the emotional content of the associated lyrics.* The latter is measured through a numerical quantity called 'emotional valence', which has a high value for emotionally positive words like love, home, life, sweet and good, and a low value for negative words like pain, death, fear, lost and sad.

One result of the study won't come as a surprise to any musician. Major triad chords tend to be associated with higher valence lyrics and minor triads with lower valences. That's always been one of the great clichés of Western popular music (though most likely an artificial and self-perpetuating one, because it's not true in other musical traditions). Unexpectedly, however, the study found that as high as major triads score, they're not the winners on the emotional valence scale. That honour goes to a much rarer chord type, known as the minor seventh. That's what you get, for example, if you play the notes A, C, E and G – which can either be taken as an A minor triad with added G, or a C major triad with added A.

Studies like this are only scratching the surface of what could potentially be done with musical data these days. You could correlate chord type, or lyrical content, with how often a song has been streamed or downloaded, and then use

* Artemy Kolchinsky et al., 'The Minor fall, the Major lift: inferring emotional valence of musical chords through lyrics', *Royal Society Open Science*, vol. 4, 11 (November 2017), https://royalsociety publishing.org/doi/10.1098/rsos.170952

the results to custom-build a sure-fire hit. Music companies, by their nature, have always been trend-watchers, but now they have the world of big data to help them. Online services like Chartmetric can provide 'real-time metrics', such as artist's radio airplay, social media engagement, streaming statistics and listener demographics. Using this information, they claim they can accurately predict which of the million-plus music artists in their database is going to make it big within the next week.

This is changing the whole face of the music industry. For one thing, by subscribing to Chartmetric-style services, placing their music directly on Spotify and making creative use of social media for marketing, artists can bypass traditional record companies altogether. More cynically, people can put their effort, not into creating great new songs per se, but in gaming the system – working out how search and streaming algorithms work, and then using that information to create exactly what those algorithms think the world is clamouring for.

It's more than a little worrying to speculate where all this is leading. If musicians can use digital analysis and computer algorithms to tell them exactly how to write a hit song, how long will it be before someone decides to cut out the middleman and let the computer write the whole song itself?

Can machines write music?

We've seen in previous chapters how various kinds of music follow clearly defined rules that can be thought of as 'algorithms' – and running algorithms is what computers do. The idea that a computer could be programmed to write music goes back a long way – all the way to Ada Lovelace, who is often characterised as 'the first computer programmer'. Born in 1815, she was the daughter of the great poet Lord Byron, who is likewise often described as 'the first rock star'. But in Byron's case, that's just a reference to his fame and popularity, and his reputation for living a wild lifestyle (one of his girlfriends described him as 'mad, bad and dangerous to know', and she meant it as a compliment).

On the other hand, Ada's claim to be a pioneer of computer programming stands on firmer ground. It's not literally true, of course, because computers in the modern sense didn't exist in those days. But she worked with Charles Babbage, who designed (but didn't build) a mechanical contraption called the Analytical Engine, which had all the basic characteristics of a programmable computer. Unsurprisingly, Babbage was the first person to write 'software' for it – but Ada came a close second. In a note published in 1843, she described a step-by-step procedure by which the Analytical Engine could perform one specific algorithm designed to evaluate a numerical sequence called the Bernoulli numbers.

At first sight this looks like a standard mathematical table, but if you look closely, you'll see that it describes a sequence of mathematical operations to be applied to certain variables, with the results stored in other variables. There's really no other word for it; it's a computer program. So Ada knew what she was talking about when, in that same year of 1843, she proposed that a machine like the Analytical Engine could be programmed to create music:

> The Analytical Engine ... is not merely adapted for *tabulating* the results of one particular function and of no other, but for developing and tabulating any function whatever ...

Number of Operation.	Nature of Operation.	Variables acted upon.	Variables receiving results.	Indication of change in the value on any Variable.	Statement of Results.	Data.					
						$^1V_1$ $\begin{smallmatrix}0\\0\\0\\1\end{smallmatrix}$ $\boxed{1}$	$^1V_2$ $\begin{smallmatrix}0\\0\\0\\2\end{smallmatrix}$ $\boxed{2}$	$^1V_3$ $\begin{smallmatrix}0\\0\\0\\4\end{smallmatrix}$ $\boxed{n}$	$^0V_4$ $\begin{smallmatrix}0\\0\\0\\0\end{smallmatrix}$ $\boxed{}$	$^0V_5$ $\begin{smallmatrix}0\\0\\0\\0\end{smallmatrix}$ $\boxed{}$	$^0V_6$ $\begin{smallmatrix}0\\0\\0\\0\end{smallmatrix}$ $\boxed{}$
1	×	$^1V_2 \times {}^1V_3$	$^1V_6, {}^1V_4, {}^1V_5$	$\begin{Bmatrix}^1V_2 = {}^1V_2 \\ ^1V_3 = {}^1V_3\end{Bmatrix}$	$= 2n$	...	2	n	$2n$	$2n$	$2n$
2	−	$^1V_4 - {}^1V_1$	$^1V_4$	$\begin{Bmatrix}^1V_4 = {}^2V_4 \\ ^1V_1 = {}^1V_1\end{Bmatrix}$	$= 2n-1$	1	...	...	$2n-1$		
3	+	$^1V_5 + {}^1V_1$	$^1V_5$	$\begin{Bmatrix}^1V_5 = {}^2V_5 \\ ^1V_1 = {}^1V_1\end{Bmatrix}$	$= 2n+1$	1	...	...	...	$2n+1$	
4	÷	$^2V_5 + {}^2V_4$	$^1V_{11}$	$\begin{Bmatrix}^2V_5 = {}^0V_5 \\ ^2V_4 = {}^0V_4\end{Bmatrix}$	$= \dfrac{2n-1}{2n+1}$	...	...	...	0	0	...
5	+	$^1V_{11} + {}^1V_{11}$	$^2V_{11}$	$\begin{Bmatrix}^1V_{11} = {}^2V_{11} \\ ^1V_2 = {}^1V_2\end{Bmatrix}$	$= \dfrac{1}{2} \cdot \dfrac{2n-1}{2n+1}$	...	2	...	...	...	
6	−	$^0V_{13} - {}^2V_{11}$	$^1V_{13}$	$\begin{Bmatrix}^2V_{11} = {}^0V_{11} \\ ^0V_{13} = {}^1V_{12}\end{Bmatrix}$	$= -\dfrac{1}{2} \cdot \dfrac{2n-1}{2n+1} = A_0$	...	...	...	...		
7	−	$^1V_3 - {}^1V_1$	$^1V_{10}$	$\begin{Bmatrix}^1V_3 = {}^1V_3 \\ ^1V_1 = {}^1V_1\end{Bmatrix}$	$= n-1 \, (= 3)$	1	...	n	...		

An excerpt from Ada Lovelace's 'computer program'.

Ada Lovelace, public domain, via Wikimedia Commons

It might act upon other things besides *number*, were objects found whose mutual fundamental relations could be expressed by those of the abstract science of operations, and which should be also susceptible of adaptations to the action of the operating notation and mechanism of the engine. Supposing, for instance, that the fundamental relations of pitched sounds in the science of harmony and of musical composition were susceptible of such expression and adaptations, the engine might compose elaborate and scientific pieces of music of any degree of complexity or extent.*

That's amazingly forward-looking, and it's only in relatively recent years that we've got anywhere near Ada's vision. Even with the advent of the first practical computers in the 1950s, they were initially employed only in very limited ways in

---

* 'Analytical Engine', Museum of Imaginary Musical Instruments, http://imaginaryinstruments.org/lovelace-analytical-engine/

a musical context. We saw in Chapter 4 how John Pierce and Max Mathews used computers to generate new musical sounds and timbres, but their compositions were still written by themselves, not a computer. Moreover, they were engineers, not musicians, so this falls in the category of scientific experimentation rather than artistic creativity.

The first 'real' composer to make use of a computer in his work was Iannis Xenakis – who was born in Romania to Greek parents and spent most of his career in France. Like John Cage in America, he was interested in the idea of aleatoric music, where an element of randomness is introduced into the composition process. But Xenakis went about it in a very different way from Cage. While the latter interpreted 'randomness' in the way that you or I might, by throwing dice or consulting the *I Ching*, Xenakis took his cue from the science of stochastic processes. These still involve random events, but within a mathematical framework that can give them a much more complex overall pattern.

To start with, Xenakis tried doing the necessary calculations by hand, but quickly found they were much too difficult and time-consuming. So he was forced to turn to a computer, but only as an aid in evaluating the calculations he was going to do anyway. The computer wasn't composing the music – Xenakis was. It wasn't electronic music, either; it was written to be played by conventional ensembles of human players. For example, an early Xenakis composition called *ST/4–1,080262*, which he originally conceived in 1956, is scored for string quartet. The 'ST' stands for stochastic, '4' means there are four instruments and '1' means it's the first work in that category; 08/02/62 is the date he finally completed it, after gaining access to a computer at an IBM research centre in Paris.

Around the time Xenakis first conceived *ST/4–1*, mathematician Leonard Isaacson and chemistry professor Lejaren Hiller of the University of Illinois were programming ILLIAC 1 – the Illinois Automatic Computer – to write its own music. Or they sort of were, anyway. It didn't work the way you might imagine, with Isaacson and Hiller feeding in a single roll of paper tape, or a stack of punched cards or whatever passed for a computer program in those days, and then ILLIAC flashing and whirring away for a few minutes before blasting out its electronically synthesised composition.

What they actually did was get the computer to perform a whole series of operations, using a variety of rules and algorithms such as the Markov chain described in the previous chapter. Hiller, who was an amateur musician at the time and later became a full-time composer, then translated ILLIAC's output into musical notation that could be played by human instrumentalists. The result was the four-movement *Illiac Suite* of 1957 – scored, like Xenakis' *ST/4-1,080262*, for string quartet.

It's really interesting to compare the two works, because – for me at least – the reaction is the exact opposite. With *ST/4–1,080262*, written by a talented human composer in a novel way that happened to involve a computer, my immediate reaction at the start was: 'What is this? It's not like anything I've heard before. I don't think I'm going to like it.' But within 30 seconds the music started to draw me in, and I found myself listening very intently, and after 60 seconds it had thoroughly captivated me. Woven from ever-changing, overlapping patterns that sound nothing like any previous string quartet, the music is entrancingly effective.

The computer-composed *Illiac Suite*, on the other hand, sounds pretty decent for the first few seconds, 'almost like a real string quartet', you think to yourself. But within a minute or so it's clear that the music simply isn't going anywhere or conveying anything. In other words, it sounds exactly as if it was written by a machine.

Over 60 years after the *Illiac Suite*, the same basic problem exists. If you program a computer to write music by following a set of rules that you give it, the result may sound pretty convincing on a bar-by-bar basis, but it's going to lack meaning and substance if you listen to the whole work. You can try it yourself if you like because there are plenty of websites and apps that allow you to generate your own unique compositions at the click of a button.* Some work better than others, but they all suffer from the same problem. Within 30 seconds, you can tell it was written by a computer.

Of course, the more sophisticated the algorithm, the better the results are likely to be. One of the most impressive purely algorithmic compositions is a piece for (human-played) violin, clarinet and piano called *Hello World!*. It was produced in 2011 by a dedicated music-composing computer called Iamus, operated by the University of Málaga in Spain. Rather than a completely rule-based system, Iamus uses 'genetic algorithms' to randomly evolve a composition towards an optimal configuration in a way that emulates biological evolution. The result is very pleasant-sounding – certainly better than the *Illiac Suite* or anything the mu-tech online composer can produce – but it's still a long way short of being great music.

As the *Guardian*'s music critic Tom Service wrote, 'The material of *Hello World!* ... is so unmemorable, and the way

---

* Here's a great one called Music Laboratory that allows you to select from a huge range of musical categories and styles: http://www.mu-tech.co.jp/AcsWebE/setparam.asp

it's elaborated so workaday, that the piece leaves no distinctive impression.' He went on to make the point everyone must be thinking by now:

> If you've got a computer program of this sophistication, why bother trying to compose pieces that a human, and not a very good human at that – well, not a compositional genius anyway – could write? Why not use it to find new realms of sound, new kinds of musical ideas?*

The great thing about human musicians is that they're always doing something new – and that's what the world really wants. Remember the Wundt curve? That referred to an *individual*'s response to new music, but it applies to collective taste too. Every now and then we need a completely new sound, like Stockhausen gave us in *Kontakte* or Xenakis in *ST/4–1,080262*. But it doesn't have to be that extreme. Think of the impact The Beatles must have had on pop listeners with 'Strawberry Fields Forever' in 1967. By combining well-established elements in a way that had never been done before, they managed to create an enduring four-minute masterpiece.

You're never going to achieve that with algorithms, in the sense of clearly defined sets of rules that the computer is constrained to follow. In a sense, the whole point of great music is to break rules, or make your own, rather than follow rules someone else has given you. As paradoxical as it sounds, there is a way to program a computer that doesn't involve rule-based algorithms. Called 'artificial intelligence',

---

* Tom Service, 'Iamus's *Hello World!* – review', *Guardian*, 1 July 2012.

or AI, it works a bit like a massively scaled-down version of the human brain.

As it happens, it's a subject I have some experience of myself. My career as a research scientist wasn't what you'd call spectacular, but I did happen to co-author a couple of papers which turned out to be well ahead of the curve. The first, dating from 1985, was about giant black holes at the centres of galaxies – years before Muse made 'Supermassive Black Hole' a household phrase. The second, which is the one that's relevant here, was published in 1997 (though the work was done several years earlier) and described the use of 'artificial neural networks' to simulate human behaviour. My co-author on that one was, at the time, a student of Kevin Warwick at Reading University – the nutty professor who became known as 'Captain Cyborg' because he voluntarily had a 100-electrode silicon controller implanted in his arm.

A neural network isn't really an algorithm at all, in the sense we've used the term so far. The programmer doesn't have to give it any rules at all, which is great because – in the study I was involved in – we didn't know what the rules were. The person we were working with, a helicopter pilot, didn't even think in terms of rules. He just responded in a certain way when presented with a certain situation. All we had to do was record his actions over a large number of trials, then give all that data to our neural network. This really is more or less what its name suggests – a network of digital nodes connected to each other like the neurons inside a brain. The idea is that the network 'learns' the sort of response to give in a particular situation, which it can then extrapolate to situations it hasn't encountered before.

I can't remember how many nodes our network had, but I think it was about a dozen, arranged in three layers.

That compares rather unfavourably with a real human brain, which has about 86 billion neurons. Modern 'deep-learning' networks do a lot better, with millions of neurons arranged in a large number of layers (that's why they're called deep). They're not the only form of AI, but they've become by far the commonest, to the extent that if you see the term 'AI' in the media, it almost certainly refers to a deep-learning neural network.

It's easy to see how a deep-learning AI could end up creating more interesting music than a rule-based algorithm. At the simplest level, you could just play it hundreds of thousands, or millions, of music tracks and leave it to work things out for itself. Or you could give it some additional information, such as the date, genre and artist or composer, so it could make connections between that information and the sound of the music. You could even tell it all those dreaded 'rules' of music, but only to ensure that it knows what they are, rather than forcing it to obey them.

So where are we heading with all of this? Will AI ever be able to create music that can pass for human – or maybe even surpass it? One person who thinks that could well happen is the musician Grimes – who studied neuroscience at McGill University in Canada before taking up her current musical vocation. Speaking on a science podcast in November 2019, she said she believed that 'in the next 10 years, probably more like 20 or 30 years' AI would advance to the point where it effectively understood the arts and sciences as well as any human. After that, she said, 'they're gonna be so much better at making art than us'.

It's a thought-provoking statement, but I doubt that many people are going to agree with her. The biggest stumbling block is her use of the word 'art', because it's so closely tied to the idea of *human* creativity and *human* emotional response. A computer might inadvertently come up with a piece that some listeners regard as art – almost certainly buried among a whole heap of lesser works – but it would be humans who decided it was art, not the computer.

In any case, there's the show-stopping point Tom Service made about Iamus: why on earth use a computer to do something humans enjoy doing themselves? There's never been any shortage of talented flesh-and-blood musicians bursting with new ideas. And to many people, music is perceived as a direct one-to-one communication between the composer or songwriter and themselves. The whole point of it would be lost if it was merely a product of AI.

That's not to say there isn't a useful role for AI in music, however. Let's go back to Grimes for a moment, who's unusual among pop musicians in not slotting neatly into one genre or another. In fact, she seems to thrive on doing something different with each new song. You can therefore imagine she might enjoy having a sort of AI assistant, which could constantly throw up off-the-wall musical ideas – with new chord progressions, melodic lines, synthesiser timbres and so on. Every hundredth idea might be good enough to give her the germ of a new song – but by the time she was finished it would be a thoroughly Grimes song. The AI would simply serve as a source of inspiration. A number of mainstream (as opposed to purely experimental) albums have already been created with AI support, perhaps most notably Holly Herndon's *PROTO* from 2019. The official AllMusic reviewer was sufficiently impressed by this to award it 4.5 out of five stars.

Another point, which can be missed by people who instinctively throw up their arms in horror at any suggestion of computer music, is that not all music is art. It sounds brutal, but it's true. A lot of music is, and always has been, purely functional. The countless 'divertimentos' of the 18th century were deliberately written as background music for social gatherings and were never meant to be properly listened to. The same is true of Erik Satie's 'furniture music' and its lineal descendent, Brian Eno-style ambient music. Then there's 'elevator music' and the background music in films and TV shows. You're not even supposed to consciously hear this music – it's just there to set the mood.

An AI could easily supply that sort of mood-setting background music. OK, there's no shortage of human composers for big-budget Hollywood movies, but what about composers of music for shopping channels, instructional videos and so on? Sometimes they may require lengthy compositions that it simply isn't cost-effective for human musicians to create. And I haven't even mentioned video games yet. While the big blockbuster games may warrant movie-style soundtracks, what about all those casual puzzle and brainteaser games where people may spend fifteen or twenty minutes at a time solving a problem or working out the next move? They don't want to listen to the same 90-second loop over and over again – yet all too often that's exactly what they get.

At least one company, Amper, has picked up on this idea already. They offer an AI-based music creation platform specifically aimed at content creators who are looking for music to accompany things like videos, podcasts or games. The old solution would have meant resorting to a stock music supplier, but with Amper you're guaranteed music that's unique to you and tailored to your specific needs. Another popular AI-powered product is LANDR, an online mastering service that uses AI to decide what digital post-processing will work best on any music track it's presented with.

One way or another, it seems likely that we'll be hearing a lot more AI-generated – or AI-assisted – music in the coming years. On the other hand, the Earth may be hit by a massive solar storm that destroys the world's electrical networks, wipes out all our digital data and sends society back into the pre-industrial age. But even if all our modern technology goes down the drain, we can rest assured that music of some form will still be with us.

'Ah, but it won't be scientific music any more,' you might reply. Actually I'd dispute that. What I hope I've shown in the course of this book is that, even in the absence of synthesisers, streaming services and all our other digital technology, it's still meaningful to talk about a 'science of music'. Pythagoras and Galileo didn't have any of our modern technology, but they were able to analyse musical sounds – and the way different combinations of sounds produce harmonious or dissonant effects – based on the wavelength and frequency of the corresponding sound waves. That's science. So too are the quantifiable numerical relationships that produce danceable rhythms and catchy melodies – or, for that matter, the totally undanceable rhythms of math rock or the spiky, otherworldly melodic lines of twelve-tone music.

FURTHER RESOURCES

Books

Bartkowiak, Mathew J., (ed.), *Sounds of the Future: Essays on Music in Science Fiction Film* (Jefferson, N.C.: McFarland, 2010)

Collins, Nick, Schedel, Margaret, and Wilson, Scott, *Electronic Music* (Cambridge: Cambridge University Press, 2013)

Forte, Allan, *The Structure of Atonal Music* (New Haven: Yale University Press, 1973)

Griffiths, Paul, *Modern Music and After* (Oxford: Oxford University Press, 1995)

Levitin, Daniel, *This Is Your Brain on Music: The Science of a Human Obsession* (London: Atlantic Books, 2008)

Maconie, Robin, *Stockhausen on Music* (London: Boyars, 1989)

Maor, Eli, *Music by the Numbers: From Pythagoras to Schoenberg* (Princeton, New Jersey: Princeton University Press, 2019)

Margulis, Elizabeth Hellmuth, *The Psychology of Music: A Very Short Introduction* (Oxford: Oxford University Press, 2019)

Marshall, Steve, *Acoustics: The Art of Sound* (Glastonbury: Wooden Books, 2022)

May, Andrew, *The Science of Sci-Fi Music* (Cham, Switzerland: Springer, 2020)

Pierce, John R., *The Science of Musical Sound* (New York: Scientific American Books, 1983)

Shawn, Allen, *Arnold Schoenberg's Journey* (New York: Farrar, Straus and Giroux, 2002)

Winterson, Julia, *Rock and Pop Theory: The Essential Guide* (London: Faber Music, 2014)

Wilks, Yorick, *Artificial Intelligence: Modern Magic or Dangerous Future?* (London: Icon, 2019)

Online resources

Algorithmic Composer – http://www.algorithmiccomposer.com/
(*An interesting resource for anyone curious about how algorithmic composition works*)

EDMProd YouTube channel – https://www.youtube.com/c/EDMProd/videos
(*Instructional videos on EDM production for beginners and more advanced students*)

Frans Absil Music YouTube channel – https://www.youtube.com/c/FransAbsil/videos
(*Unconventional music analysis from a science PhD and former aerospace engineer*)

'Guide to Electronic Music Genres', YouTube – https://www.youtube.com/watch?v=81ZGPVP0hb4
(*... and there really are an awful lot of them*)

'How Pythagoras Broke Music', YouTube – https://www.youtube.com/watch?v=EdYzqLgMmgk
(*A detailed look at the pros and cons of Pythagoras' theory of music*)

'How To Make a #1 Song – Without Talent', YouTube – https://www.youtube.com/watch?v=kKnxcvNqrCo
(*An amusing video that almost lives up to its billing*)

Jacob Collier's YouTube channel – https://www.youtube.com/user/jacobcolliermusic/videos
(*A fascinating insight into how a contemporary singer-songwriter works*)

MuseNet – https://openai.com/blog/musenet/
(*An online AI that can create new music in a variety of styles, starting from a few notes of an existing work*)

Music Map – https://www.music-map.com/
(*Type in your favourite artist or composer, and see how they relate to others in similar genres*)

'The Science of Music' YouTube playlist – https://www.youtube.com/
    playlist?list=PLvQfFzOInEpTjvazcXJE9ykgCuxzatY_v
(Electronic compositions created as spinoffs from this book)

'Practice Exercises' by ToneSavvy – https://tonesavvy.com/
    music-practice-exercises/
(Interactive music theory tests for anyone brave enough to try them)

'Return of the Monster from the Id', *Between the Ears*, BBC Radio 3 –
    https://www.bbc.co.uk/programmes/b03bfdkj
(A radio programme about the making of the Forbidden Planet *soundtrack in
the 1950s)*

Singing Voice Synthesis System – http://sinsy.jp
*(An online app that allows you to input a melody line and lyrics to produce
synthesised vocals)*

'Sonification of Celestial Data', Florida International University –
    https://faculty.fiu.edu/~fterenzi/research/
(A selection of Fiorella Terenzi's astronomical sonifications)

'Stockhausen: Sounds in Space' – http://stockhausenspace.blogspot.
    com/
*(A blog that does wonders clarifying the logic behind some of Stockhausen's
more baffling works)*

'Synthesizers, as Digested by a Classical Musician', YouTube –
    https://www.youtube.com/watch?v=i3Ag5pwNSiM
*(Nahre Sol, a pianist of the social media generation, explores the world of
synthesisers)*

'Synthesizer basics overview', Logic Pro User Guide –
    https://support.apple.com/en-gb/guide/logicpro/lgsidcd8a98b/
    mac
*(An appendix to the Logic Pro User Guide containing a nice clear explanation
of how synthesisers work)*

Theory Tab Database – https://www.hooktheory.com/theorytab
(Find out what key or mode a song is in, and what chords it uses)

'The Tonnetz diagram and Neo-Riemannian Theory', Frans
    Absil – https://www.fransabsil.nl/htm/tonnetz_riemannian_
    transformations.htm
(Learn how to navigate from one chord to another, Hollywood-composer style)

'Uncommon Time', TV Tropes – https://tvtropes.org/pmwiki/
pmwiki.php/Main/UncommonTime
(A surprisingly long list of songs with irregular time signatures)

Software

Audacity – https://www.audacityteam.org/
(Free, open-source audio processing software)

Chord ai – https://www.chordai.net/
(An AI-based app that identifies the chords in a song in real time)

Computer Music Magazine – https://pocketmags.com/
computer-music-magazine
*(Buy any single issue and get a whole suite of free DAW plugins and other
electronic goodies)*

MuseScore – https://musescore.org/en/download
*(If you know traditional music notation, this software lets you create and play
your own scores)*

OpenMusic – https://openmusic-project.github.io/openmusic/
(A great free way to get into algorithmic programming)

Waveform Free – https://www.tracktion.com/products/
waveform-free
(A completely free, fully functional digital audio workstation)

CHRONOLOGICAL PLAYLIST

This isn't an exhaustive list of all the music referred to in the book, just pieces that are good illustrations of various topics that have been discussed.

1785	Wolfgang Amadeus Mozart	String Quartet No. 19 ('Dissonance'), first movement
1813	Ludwig van Beethoven	'Battle Symphony' ('Wellington's Victory')
1827	Franz Schubert	'Frühlingstraum' from *Winterreise*
1830	Hector Berlioz	*Symphonie Fantastique*, fifth movement
1854	Franz Liszt	'Faust Symphony', first movement
1865	Richard Wagner	*Tristan and Isolde*, prelude
1876	Richard Wagner	*Siegfried*, Act 3
1885	Arthur Sullivan	'Miya Sama' from *The Mikado*
1885	Franz Liszt	'Bagatelle Sans Tonalité'
1912	Arnold Schoenberg	*Pierrot Lunaire*
1935	Alban Berg	Violin Concerto
1948	Arnold Schoenberg	*A Survivor from Warsaw*
1949	Olivier Messiaen	*Turangalîla-Symphonie*
1953	John Cage	*Williams Mix*
1956	Bebe and Louis Barron	*Forbidden Planet* soundtrack
1957	Milton Babbitt	'All Set'
1958	Edgard Varèse	'Poème Électronique'
1959	Miles Davis	'So What'

1960	Karlheinz Stockhausen	*Kontakte* (preferably the version *without* a piano)
1962	Iannis Xenakis	*ST/4-1,080262*
1963	Delia Derbyshire	*Doctor Who* theme (original version)
1966	Igor Stravinsky	'The Owl and the Pussy-Cat'
1967	The Beatles	'Strawberry Fields Forever'
1967	Frank Zappa	'Brown Shoes Don't Make It'
1967	The Doors	*Strange Days*
1968	The Jimi Hendrix Experience	*Electric Ladyland*
1968	Jerry Goldsmith	*Planet of the Apes* soundtrack
1968	Wendy Carlos	*Switched-On Bach*
1969	Black Sabbath	'Black Sabbath'
1969	Rosemary Brown	'Grübelei' (as by 'Franz Liszt')
1970	Karlheinz Stockhausen	*Mantra*
1970	Frank Zappa	'Toads of the Short Forest'
1974	Tangerine Dream	*Phaedra*
1974	Isao Tomita	*Snowflakes Are Dancing*
1974	Kraftwerk	'Autobahn'
1975	Brian Eno	*Discreet Music*
1975	Peter Maxwell Davies	'Ave Maris Stella'
1980	John Williams	'The Imperial March' from *The Empire Strikes Back*
1982	Thomas Dolby	'She Blinded Me With Science'
1984	Art of Noise	'Close (To the Edit)'
1984	Frankie Goes to Hollywood	*Welcome to the Pleasuredome*
1984	Iron Maiden	'Rime of the Ancient Mariner'
1988	Pet Shop Boys	*Introspective*
1992	The Prodigy	*Experience*
1993	Aphex Twin	*Surfing on Sine Waves* (as by Polygon Window)
1995	William Orbit	*Pieces in a Modern Style*
2000	Radiohead	*Kid A*
2001	Gorillaz	'5/4'
2008	Meshuggah	'Dancers to a Discordant System'
2011	Björk	*Biophilia*
2012	Mickey Hart	*Mysterium Tremendum*
2014	Taylor Swift	'Shake It Off'
2016	Emily Howard	*Magnetite*
2017	Ed Sheeran	'Shape of You'
2019	Holly Herndon	*PROTO*
2020	Grimes	*Miss Anthropocene*
2020	Jacob Collier	*Djesse Vol. 3*

INDEX